LOUIS VUITTON

A PERFUME ATLAS

LOUIS VUITTON

A PERFUME ATLAS

Texts by Jacques Cavallier-Belletrud and Lionel Paillès

Illustrations by Aurore de la Morinerie

Photographs by Sébastien Zanella

A Small Geography of Scents

Perfume is a journey, intimate and motionless at first, an illusion that takes you by the hand to lead you somewhere else, to another emotion, another dimension, a completely different space-time.

Having been born in Grasse, I know better than anybody that the raw materials bestowed on a perfumer come from all over the world. Some of them grow just a few hundred metres from my home in Provence, but others – the vital ones – are found on distant shores with enticing names: Somaliland, Nosy Be, the Moluccas, Guilin...

A fragrance starts off as flowers, leaves, seeds, roots or rhizomes generously offered to us by nature, and we try to extract their scents as faithfully as possible. These scented plants flourish in distinctive soils, in various climates, giving them their own specific olfactory characteristics. This shows that the concept of 'terroir' is not an idle fancy but rather something alive and tangible, a botanical, spatial and cultural reality.

I have come to realize that, when all is said and done, I have learned about geography through raw materials. Patchouli is traditionally cultivated in Indonesia; I found that out very early on, by asking, 'What country could produce such a fragrant and mysterious ingredient?' The same went for myrrh, benzoin, incense and osmanthus. I must have been fourteen when I stuck a blue Dymo label reading 'osmanthus' on a flacon – it symbolized for me the other side of the world, unknown lands that I would never visit. Learning about scents suddenly became an exhilarating game, and the perfumer's palette became the most amazing planisphere.

Magellan, Marco Polo and Captain Cook have taught us that a map is not merely a tool of knowledge but also a mental representation, a personal vision of reality. In this atlas of scents, I would be inclined to place Grasse at the centre – not out of any undue bias but because all the most important fragrant plants on this planet have been nurtured, transformed and extracted there. It could be said that they fully came into their own in Grasse. Leaving aside my hometown, however, almost the entire planet is represented in this atlas – most of the continents, as well as a large number of countries, both near and far.

When I began to work for Louis Vuitton, the luggage company that has made travel its DNA, I was naturally keen to pay homage to these essences and absolutes, as well as the perfume plants expressed through them – not to mention the men and women who grow them, day in, day out, and the great civilizations that brought them to life, sometimes thousands of years ago.

The perfume plants that appear one after the other in the following pages, such as key lime, vetiver and oud, now strike me as a poetic vision that offers a way to see the world differently, in a more democratic and egalitarian light, where everything has its place and a role to play, where the wealthiest continents are not necessarily the ones that produce the rarest scents. This vision upholds the reality that most of these plants no longer grow in the lands where they originated. They have been transplanted and acclimatized to foreign settings, where they have achieved their maximum expression. There is no doubt that the geography of scents is changing. This book bears witness to those parts of the world that have come to the fore in the production of raw materials, and we are indebted to them for the finest ingredients on offer today. In five or ten years' time, others will surely have taken their place. The Spice Route is now nothing more than a myth that serves as a basis for adventure novels. New routes have long since been opened up, revealing new paradises of fragrances.

Whether you are an inveterate globetrotter or a resolute homebody, this atlas will bring alive the scents of the world, from Calabria to Guatemala, from China to the United States, where the essences that have been leading me by the nose for so many years are found. I hope this book will inspire each and every reader to embark on the most extraordinary of voyages, the one provided by perfume, which wafts from one person's skin into another's heart.

Jacques Cavallier-Belletrud

To Claire, Camille and Clémence, the women in my life.

The possibility of an island

The thing that first strikes the eye about the island of Réunion is that nature is queen: there are mountains everywhere, with strikingly shaped peaks, as well as ravines, waterfalls and forests. If there were ever a land that was the natural projection of a perfumer's senses, it is this one. Its splendour is reflected in every living thing that makes its home there.

This green paradise is home to high-quality perfume ingredients. Prime among these is geranium Bourbon (which should strictly be called geranium Bourbon of Réunion, because the name Bourbon has also been extended to Madagascar): an exceptional product that lingers twice as long as Egyptian geranium, it is grown in the western highlands of the island at altitudes of over 1,000 metres.

Château Pilsen 50/50 2RP

Another is baie rose, also known as pink peppercorn, although it is not a true member of the pepper family; wonderfully, it can even be found growing wild by the roadside. According to the island's people, it is Piton de la Fournaise – the 'kind' volcano – that gives the land its richness. Travelling in Réunion feels like a journey back to the dawn of history.

The fruits of Calabria

A magical moment that I never tire of, even though I've seen it many times, is the sight of the Calabrian landscape of Italy dotted with bright bergamot fruits in January. Bergamot thrives throughout the southernmost parts of the province of Reggio di Calabria, especially on the Ionian coast. Growing conditions there are ideal: the humidity levels, the light, the air currents that offer protection from winter cold and summer heatwaves.

It's in this land, and this land alone, that the fruit attains its peak. There's such sensuality in the act of crushing a bergamot leaf in your hand, releasing its scent, or piercing the skin of the fruit to smell the zest. What strikes me in Calabria was that every view is an extraordinary tableau that combines the blue of the sea and the green of the citrus orchards and olive trees. The countryside is absolutely beautiful yet harsh at the same time. This contrast gives a powerful boost to the imagination of a perfumer like me.

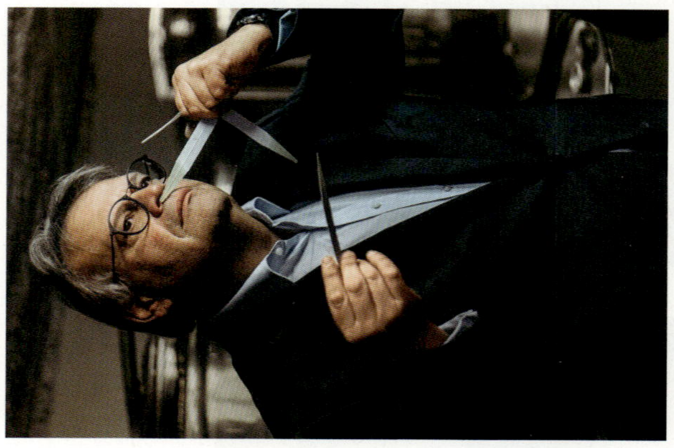

Maya gold

This journey took us first to Guatemala, nicknamed the ‘land of eternal spring’ because the weather is warm all year round. This is where they cultivate cardamom, a wonderful spice that grows in humid mountain areas in the shade of coffee bushes. I remember the huge volcanoes that seemed to take us thousands of years back in time, and the little white towns perched on rocky peaks.

Three hours away by road, El Salvador is the secret home of Peru balsam. After travelling for hours down an almost impenetrable track, you're surrounded by dark forest. The sight is indescribable: in the ancient trees, 25 metres in the air, men peel off the bark and burn the bare trunk with flaming torches to make it exude the precious resin, just as the Maya people once did. A woman also prepared a Maya chocolate drink for me, made from cocoa, cinnamon and chilli.

Every moment I spent there could have inspired a fragrance. Once again, my long-standing affection for ingredients deepened into love when I got to know the plants involved and the people who grow them.

Africa

Cocoa
Ivory Coast

*'There is a kind of magic in that soothing childhood memory of opening a tin of cocoa powder – and when cocoa meets patchouli, it's a veritable enchantment!'**

The cocoa bean originated in Central and South America. In pre-Columbian times, the Aztecs used it as currency and considered it a gift from the gods. It was the main ingredient in *xocoatl*, a frothy, bitter drink that was consumed in religious ceremonies honouring the great god of the forest.

Cocoa subsequently made its way to Africa, albeit centuries later. The cocoa tree was not introduced into Ivory Coast until 1880, at the instigation of Arthur Verdier and Amedée Bretignère, traders from La Rochelle who were instrumental in the creation of a French protectorate in the area. Cocoa acclimatized itself so quickly that it now seems almost indigenous, for it was perfectly suited to the humid tropical forests that were barely penetrable by sunlight. These days, the epicentre of cocoa production is in the Nawa region to the southwest of Ivory Coast. As the dry and rainy seasons are clearly differentiated, two annual harvests are possible: the main one extends from October to March, with a low season running from April to August.

The cocoa fruit, which grows from the tree's branches as well as directly from its trunk, is green and oval, with a thick, rough shell concealing a fleshy pulp. A single fruit can weigh up to one kilo. It is ripe when it takes on a bright yellowy-orange tinge and the beans can be heard rattling inside. In fact, this sound signals the start of the harvest. The fruits are chopped off the tree with a machete, although those growing furthest from the ground have to be cut down with a special sharp-edged tool attached to a long pole. Considerable care has to be taken to avoid any incisions in the tree itself, as these could open the way for a parasitical fungus. Experienced pickers become so skilled that they can dislodge up to 1,500 fruits in one day.

The shells of the fruits are split into two on the ground; the cocoa beans (fifteen to twenty on average) are removed and laid out on banana leaves. They are then piled up indoors, in several layers, for two to eight days, depending on the variety. This stage is decisive, as it is here that the beans start to develop their aromatic potential. They are roasted at 100–150 °C, leaving them thoroughly dried and their flavour at its fullest. Finally, the beans are ground prior to the extraction process. The absolute, the quintessence of cocoa's odiferous principle, is powdery and airy, with just the right touch of bitterness. Cocoa constitutes one of the few natural gourmand (edible) notes available to perfumers, who have been quick to take advantage of it: it can be combined with patchouli or more synthetic gustatory notes, to achieve a degree of soulfulness and naturalness. The ultimate aim is to recreate on the skin the same emotion conveyed by a square of chocolate – the expression of a moment of happiness as simple as it is addictive.

* *The quotes at the beginning of the texts on raw materials are the words of Jacques Cavallier-Belletrud.*

Genus and species: Theobroma cacao
Family: Malvaceae
Type: Small, evergreen tropical tree (4–10 m)
Main growing areas: Ivory Coast, Ghana, Ecuador
Harvest: Mainly October–March; secondary harvest April–August
Part used: Bean
Treatment: Extraction of ground beans with volatile solvents (absolute)
Olfactive family: Gourmand
Facets: Amber, balsamic, gourmand, powdery, sweet, coumarin, animal, liquorous

Incense
Somaliland

'Whenever I smell this fresh, perfumed, aldehyde resin, I am immediately transported to the Mount of Olives or a Shinto sanctuary.'

The word 'incense' usually conjures up images of sticks burning in temples or churches, but rather than authentic incense, these tend to be mixtures of oliban resin (true incense) with myrrh, pine, cinnamon or sage. In these circumstances, it is difficult to appreciate the real smell of incense; for that, it is necessary to go back to the botanical source. Long before the invention of incense sticks, there was a small, twisted tree with tangled branches, with no apparent charm. There are around thirty species of incense tree, but it is *Boswellia carterii* that produces the strongest smelling resin of all and is thus the favourite of the perfume industry. It is considered by the peoples living around it a gift from God and a compensation for limited material resources – for it grows only in an arid lunar landscape exposed to fierce sunshine and suffocating heat, 1,000 metres above sea level, in Somaliland (a self-proclaimed republic in the Horn of Africa that has been campaigning for recognition by the UN since 1991). Even finding one of these trees is a considerable achievement, as they are few and far between, but the effort is rewarded by the precious nectar concealed within its whitish trunk.

An ancient method of tree tapping is still used to reach it. Incisions 10 centimetres long are made with a *mingaaf* (traditional knife). The first resin, of mediocre quality, starts to seep out after fifteen to forty days. As it dries in the air, this substance, known as oliban, turns into the highly prized 'tears' of incense. At each subsequent collection, the tree tapper makes the incision bigger, and the next resin that appears, over the course of several months, is of far better quality. After a third harvest, the tree is allowed to 'rest' for three years. Connoisseurs prefer resin that is gathered in autumn as the result of incisions made in summer. This is white incense, in contrast to red incense. The output of resin can vary enormously from one tree to the next – while some may yield large amounts, others may provide only a few grams. A tree is considered 'generous' when it produces three to ten kilos of resin at each harvest.

In Somaliland, as elsewhere, tapping for incense is carried out by nomadic herdsmen who combine this activity with animal husbandry. They are the repository of the traditional skills involved in collecting the finest resin without ever damaging a tree.

Incense has lost nothing of its enormous evocative power: its scent still sends us back in time almost immediately – to the earliest civilizations, Ancient Egypt, the Queen of Sheba, the Three Magi, Arabia Felix, caravans across the desert...

Perfumers have put this long history to good use. They are fortunate to have at their disposal two products with complementary inflections that bring together all the subtleties of incense: on the one hand, an essential oil with fresh, spicy, citrus notes; on the other, a resinoid that provides a warm, balsamic fragrance. This complexity makes incense one of the few raw materials capable of expressing two seemingly contradictory emotions: light and darkness, heat and cold, yin and yang, East and West.

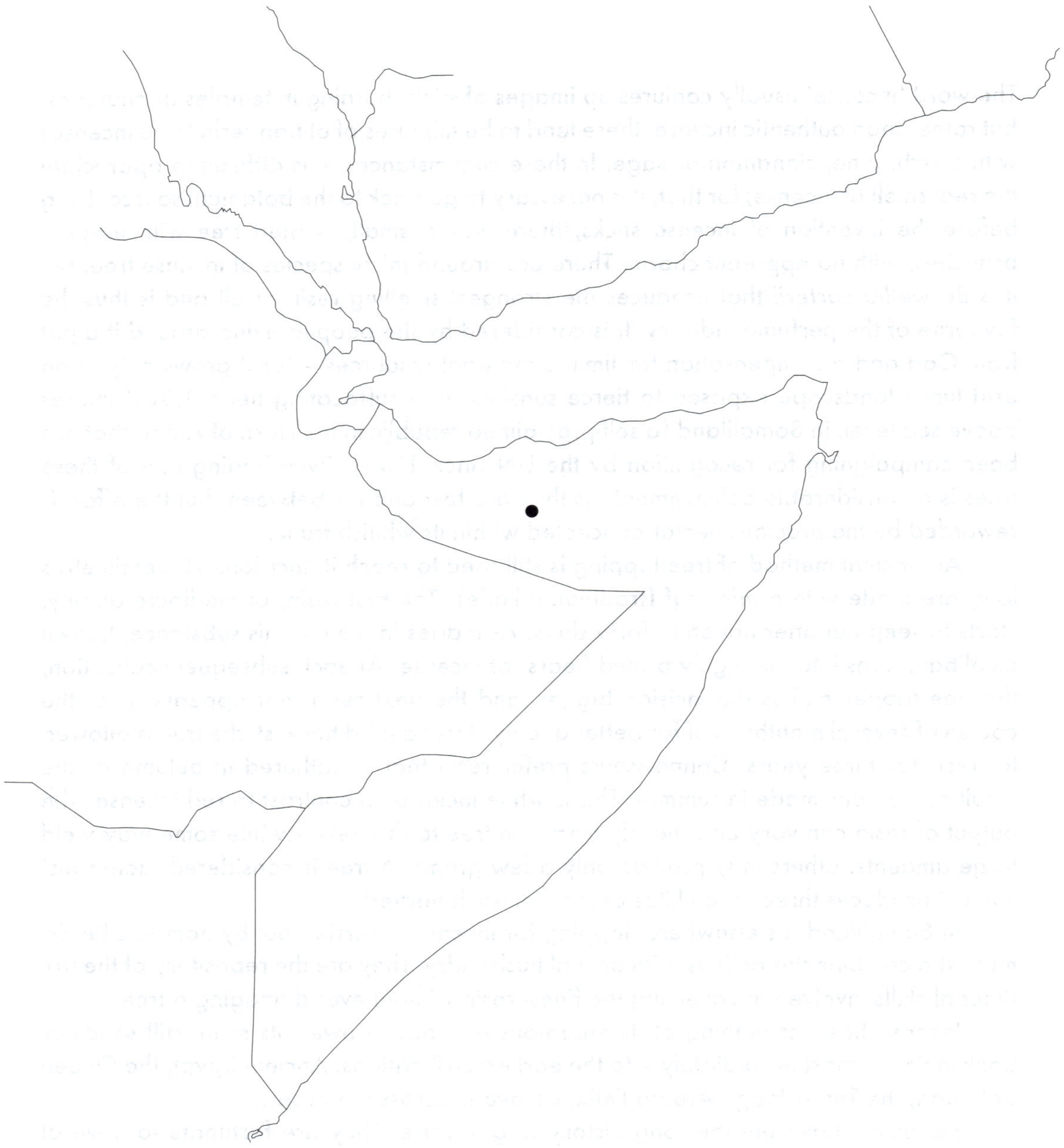

Genus and species: Boswellia carterii
Family: Burseraceae
Type: Deciduous shrub (1–2 m)
Main growing areas: Somaliland, Oman, Saudi Arabia, Ethiopia
Harvest: September–November
Part used: Gum resin
Treatment: Hydrodistillation (essence)
Olfactive family: Woody
Facets: Resinous, aromatic, peppery, camphorous, smoky

Geranium rosat
Réunion

'The geranium rosat is completely androgynous and can provide both a "clean" aromatic freshness and a natural, green florality.'

This small, highly fragrant bush so beloved of perfumers is incorrectly called a 'geranium' because of its vague resemblance to the decorative plant that brightens up our balconies and gardens, but in fact it is a pelargonium – from the Greek word *pelargos*, meaning 'stork', in reference to the shape of the fruit, which is slightly reminiscent of a stork's beak.

Many cultivars of geranium rosat have been selected and exploited for their perfume in several parts of the world. The variety known as 'Bourbon', which grows in the extraordinary garden constituted by the island of Réunion, is a real aromatic treasure and probably the finest of all: sweet and intoxicating, with a strong rose, floral scent.

The fragrant plantations are concentrated in the west of the island, around the quiet, medium-altitude village of Petite-France, considered by the locals as the 'geranium capital'. In March and April, the fields explode in a riot of pinks and purples. Once its flowers drop off, the plant becomes increasingly robust as it absorbs the intense heat and strong sunshine required to reach full maturity. The harvest follows the end of this phase. The plants' stalks are cut off with a sickle around 15 centimetres from the ground and then left in the field to wilt (it is useless to distil geranium rosat immediately, in contrast to related plants such as rose and jasmine). The precious essence – that fragrant juice that combines so well with, for example, the Damask rose – is concealed in the down under the leaves. By harvest time, the foliage is dense and compact, as the plant seems determined to mark its territory with greenness, dominate the space around it and unleash its power to all and sundry. The geranium at the start of the season is particularly green, while at the end of the season it is richer in citronellol and markedly fresher. These different qualities are brought together by the communelle, the final mixture of the various harvests.

Once gathered, the leaves are immersed in a still heated over a wood fire for at least four hours, with the fire kept ablaze manually throughout the process. Some 300 to 400 kilos of leaves are needed to obtain 0.5–1 litre of essential oil.

Probably because it shares 30 per cent of its components with the rose itself, geranium rosat is sometimes called 'the poor man's rose'. Nothing could be further from the truth, however, as, like any other raw material, it chooses neither masculine nor feminine gender. Nor is it a kind of 'sub-rose', as could be suspected from the old-fashioned custom of redistilling essential oil of geranium with rose to obtain an 'essence of geranium on rose'. The geranium rosat has its own highly distinctive olfactory properties that are original and rich, a match for all comers. This is evident as soon as the lobed leaves and stalks are unloaded into the still prior to distillation, in just the same way they were a century ago. The scent of peppermint, lemon and green apple permeates the air, and you forget everything you have ever heard about the supposed simplicity of this plant, which is as hardy as it is magnificent.

Genus and species: Pelargonium x asperum
Family: Geraniaceae
Type: Evergreen shrub (50–80 cm)
Main growing areas: Réunion, Madagascar
Harvest: June–July
Parts used: Leaves and stems
Treatment: Steam distillation (essence) and extraction with volatile solvents (absolute)
Olfactive family: Flowery
Facets: Rosy, minty, spicy-peppery, herbaceous

Ginger
Nigeria

'This tropical plant discovered by Marco Polo combines effervescence with a spicy sensuality and a comforting fullness that recalls candied ginger.'

Although the aphrodisiac properties of the tonifying, citrusy root *Zingiber officinale* have only ever really convinced the Chinese (and maybe Louis XV, who used to secrete it around his apartments before receiving his mistress Madame du Barry), this medicinal rhizome has long been popular with perfumers for quite other reasons, as from the very first waft it evokes a touch of insolence and the sparkle of youth.

Originally from China and southern India – the word 'ginger' is derived from the Sanskrit *shringavera*, meaning 'in the form of antlers', in reference to the young shoots emerging from the rhizome – this herbaceous plant was first discovered around 5,000 years ago. It requires a tropical or subtropical climate with a constant humidity level, abundant rainfall and a mean temperature of over 21 °C. These conditions have thus allowed it to flourish in China, India and central and southwestern Nigeria.

The ginger plant, which resembles a reed, has slender stalks that can reach a length of two metres, and which bear fragrant lanceolate leaves. Its greatest treasure is hidden, however, in its branching underground root – that is, its plump, gnarled rhizome – which is widely used in Asian cooking for its spicy, aromatic flavour. The rhizome's skin is particularly rich in complex fragrant substances.

Around nine months after ginger is planted, its green aerial parts begin to dry and fall off, indicating that its essential oil has descended into the roots: harvest time is between December and January. Increasingly the harvest takes place as early as possible, to ensure that the ginger is tender and not too fibrous before it is treated.

Ginger is renowned for its healthy, detoxifying qualities, as well as its antiseptic and antibacterial properties (hence its elevated status among the plants used in traditional Chinese medicine). Nevertheless, it has always had a few detractors in the perfume world, on account of a facet described as 'soapy' that was systematically found in the essential oil obtained via hydrodistillation and steam distillation. The rhizome was traditionally brushed, cleaned, stripped of any rootlets and then submerged in water to make it easier to peel, before being dried for two to three weeks. A recent development involving the treatment of fresh ginger roots with supercritical CO_2 – two days after the harvest at most – has completely changed the situation. The resulting scent is far nobler and has spurred renewed interest in this knobbly rhizome – to such an extent that nowadays it is almost approached as a brand-new ingredient.

Ginger is universally acclaimed in perfumery thanks to its duality: its bracing, effervescent, juicy, textured character is obviously familiar, but it also harbours a less well-known, smoother and warmer facet, somewhere between clove and nutmeg, which adds a great deal of depth and sensuality to a composition.

Genus and species: Zingiber officinale
Family: Zingiberaceae
Type: Herbaceous tropical plant (0.9–1.3 m)
Main growing areas: Nigeria, Sierra Leone, Jamaica, India, China (province of Shandong)
Harvest: December–January
Part used: Rhizome
Treatment: Hydrodistillation or steam distillation (essential oil) and supercritical CO_2 extract obtained from fresh roots (essence)
Olfactive family: Spicy
Facets: Peppery, lemony, rosy, woody, hot, vervain

Myrrh
Somaliland

'This resin, dating back thousands of years, generously endows perfume with warmth, texture, volume and a surplus of richness and authenticity.'

Smelling myrrh is like opening the book of humanity. It is mentioned in all the holy texts, from the Bible and the Koran to the Hindu Vedas. The Ancient Egyptians used this fragrant resin to make the sacred perfume *kyphi* and embalm the corpses of the pharaohs. In Greek mythology, Myrrha fell in love with her father, Cinyras, the king of Cyprus, and fled to Arabia in shame. The gods took pity on her and transformed her into a tree, whose bark split to give birth to her son, Adonis. According to this legend, the drops of resin that ooze from the tree are the young woman's tears, which have not ceased to flow to this day. Finally, it should not be forgotten that, in the Bible, myrrh – a symbol of the Holy Infant's human nature and mortality (keep in mind its use in the embalming process) – was one of the splendid gifts brought to Jesus by the Three Kings, specifically Balthazar, who came from Africa.

It must not be thought, however, that myrrh belongs exclusively to mythology. Its voluptuousness is very much a reality in Somaliland, the cradle of humankind. In this autonomous territory situated to the north of Somalia in the Horn of Africa (also home to the incense tree), a small, solitary, wild tree, its twisted branches bristling with thorns, grows as if by magic on arid savannahs pummelled by the sun. At the end of the dry season, it is swathed in orangey-red flowers, while its trunk and main branches are covered with swollen knots, out of which myrrh flows quite naturally, in small, whitish 'tears'. These are collected a few weeks later, once they have dried and darkened to first amber and then a reddish-brown colour. This spontaneous secretion, known as stacte, contains the most fragrant and exquisite myrrh. More sap is released from the tree twice a year by means of incisions, although this forced exudation seems to result in a resin of poorer quality. As a general rule, tears of myrrh are the size of a large hazelnut but they can sometimes be as large as an egg.

In Somaliland, collecting myrrh goes hand in hand with rearing goats and sheep. While women and children take the animals to pasture, men gather the resin by hand. This is no easy task as the tree is equipped with formidable thorns and the gum is difficult to dislodge, and the only tools usually available are stones, or just branches.

Myrrh extract develops a smell that is sweet (caramel), spicy, aniseed and pleasantly medicinal. In expert hands, it can also suggest bitterness (its name comes from *murr*, meaning 'bitter' in Arabic). This extract is a faithful reproduction of the precious resin, capable of bestowing on skin a warm, extremely sensual caress, sweeter than honey – recalling the words of the biblical Song of Songs: 'A bundle of myrrh is my well-beloved unto me; he shall lie all night betwixt my breasts.'

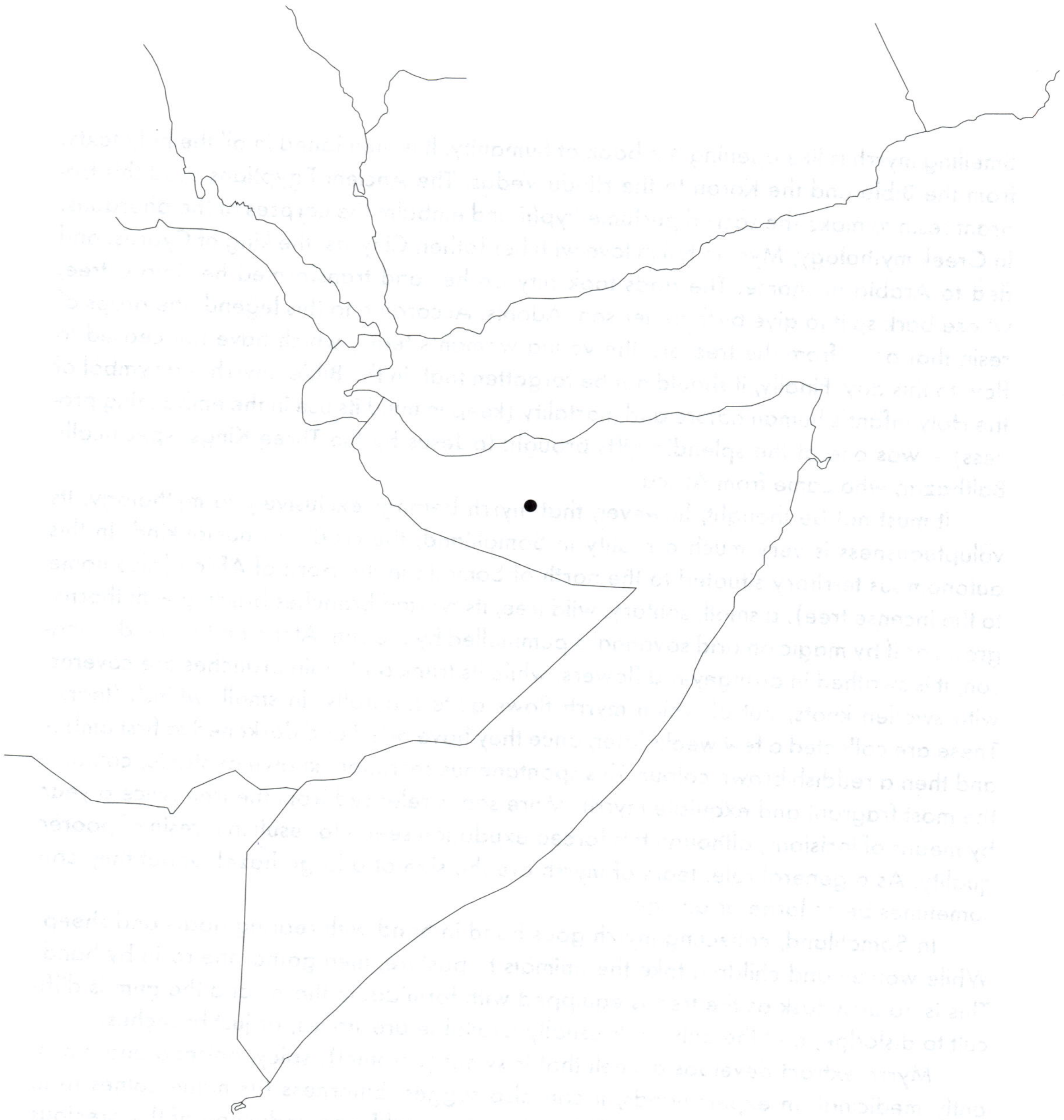

Genus and species: Commiphora myrrha
Family: Burseraceae
Type: Thorny, bushy tree (2–4 m)
Main growing areas: Somaliland, Arabian Peninsula
Harvest: October–January
Part used: Resinous exudate
Treatment: Steam distillation (essential oil) and extraction with alcohol (resinoid); supercritical CO_2 extraction also possible
Olfactive family: Amber
Facets: Balsamic, woody, sweet (caramel), warm, liquorice, medicinal

Neroli
Tunisia

'All the strength of the South is expressed in the sophisticated freshness of neroli, which conjures up childhood as well as being the most universal note that I know, the gateway to all perfumes.'

Neroli owes its charming name to Marie-Anne de La Trémoille, Duchess of Bracciano and Princess of Nerola, who, in the seventeenth century, christened this essence that she so adored after the village of Nerola on the outskirts of Rome, which belonged to her husband, Prince Flavio Orsini. She promoted neroli enthusiastically, using it to perfume her bath water, her gloves and even the sumptuous wall hangings in her Palazzo Pasquino in Rome.

'Neroli' denotes the essential oil obtained from the flowers of the bitter orange. This tree generously provides perfumers with an array of fragrant wonders, as the fruit, flowers and leaves can all be used. Each part produces an extract with distinctive tonalities: the essence of the bitter fruit, the absolute of the blossom, the essential oil and absolute of the leaves (known as petitgrain) and, of course, neroli essential oil.

For a long time, the cultivation of the bitter orange was dominated by Grasse, until the elements intervened, to disastrous effect. The terrible freeze of 1956 wiped out most of the orange groves in the region, to the benefit of plantations in Morocco, Egypt and, above all, Tunisia, where neroli essence – or, more exactly, the floral water left behind after distillation – is used to perfume the country's famous gazelle horns (pastries with a soft filling). The bitter orange tree has proved ideally suited to the mild weather of the southern Mediterranean coastline, where it can survive for several hundred years.

In the Nabeul region of northeastern Tunisia, the bitter orange blossom is a veritable 'white gold' that provides a living for much of the local community, and not only orange groves but also private gardens and village squares are awash with it until mid-April. The flowers grow in bunches and they are still picked by hand – in the morning, once the dew has evaporated. At the start of the season, the harvest is largely focused on buds, which provide greater amounts of essential oil; open flowers are generally reserved for the extraction of a concrete (a waxy mass used in the production of absolutes). Pickers usually work in groups of three or four, with the youngest member climbing a stepladder to reach the top of the tree, while the remainder gather the flowers on the lower branches. These precious treasures are then spread out on tarpaulins under the foliage. At this stage, it is vital to ensure that no twigs get mixed up with the petals, as they could add an unpleasant sourness to the essential oil.

In Tunisia, neroli is obtained via steam distillation of the fresh flowers in the twelve hours following the harvest. The amber-yellow essence develops a magnificent floral fragrance that is zesty and much lighter than the absolute of orange blossom. Neroli is one of the freshest elements derived from the bitter orange tree, along with petitgrain essence. It has therefore long been used to imbue eau de Cologne with an unpretentious nobility. The art of the perfumer is sometimes compared to that of a musician; if neroli were a melody, it would be *A Little Night Music*.

Genus and species: Citrus x aurantium
Family: Rutaceae
Type: Evergreen shrub (2–3 m)
Main growing areas: Tunisia, Morocco, Egypt
Harvest: March–April
Part used: Flower
Treatment: Hydrodistillation and steam distillation (essential oil)
Olfactive family: Flowery
Facets: Floral, fresh, lively, green, citrusy

Rosemary
Morocco

'This beautiful rustic plant, an emblem of the Mediterranean natural world, brings warmth and sensuality, over and above its vivacity.'

This plant has acquired the nicknames 'herb of crowns', 'herb of troubadours' and 'incense burner' (as it smells similar to incense), but it is 'rosemary' (from the Latin *ros marinus*, meaning 'dew of the sea') that most effectively conveys its almost compulsive need to grow on a seashore.

Rosemary belongs to the same botanical family as lavender, thyme, mint and lemon balm. It thrives on sunshine, preferring mild winters, and flourishes easily in the wild. The rustic species flourishes in poor soils, growing spontaneously on heaths, scrublands and stony ground around the entire Mediterranean Basin.

For anyone hiking in the eastern part of Morocco, a few distinguishing signs make it easy to spot rosemary, starting with its azure-blue flowers with two sharply defined lips, clustered in the axils of the leaves rather than at the ends of the branches. Above all, any walker undeterred by the intense heat will quickly notice its invigorating camphorous fragrance, reminiscent of the captivating smell of eucalyptus, wafted aloft by even the slightest breeze.

The rosemary harvest takes place in spring, when the plant reaches its olfactory peak. The presence of flowers gives the starting signal, as any raw materials collected before their appearance would be liable to produce essential oil with a bland smell, dull and flat. The stems, complete with their evergreen, needle-like leaves, are cut off with a sickle or large shears, or by mechanical means, depending on the region. Whatever method is used, however, only the upper part of the plants is removed and great pains are taken to avoid uprooting any of them.

It is crucial to dry the harvest in the shade, in ventilated facilities, before it can be processed. Essential oil of rosemary is a deep yellow colour and it is obtained via steam distillation of the flowery stems. As for the scents that it emits, these are fairly complex: fresh, herbaceous, camphorous, highly aromatic and, a little later on in its evolution, even woody. The latter note recalls incense, albeit with a slightly animal inflection. Contrary to received opinion, rosemary has other uses apart from flavouring grilled meat, infusions and honey, or stimulating students' brain cells and improving their memory. It is one of the oldest raw materials known to perfumery, as if the primary purpose of its existence was to provide pleasure. It should not be forgotten that the first recorded alcohol-based perfume, Queen of Hungary water – a legendary fourteenth-century elixir that would later become the favourite of both Madame de Sévigné and Madame de Maintenon – is a decoction of rosemary branches in bloom in a small amount of alcohol. Rosemary's long record of supplying its dynamic, aromatic inflections to 'historic' perfumes speaks for itself, but any attempt to reduce it to this glorious past is to overlook the huge potential that its generous, rustic naturalness still has to offer to modern perfumery.

Genus and species: Salvia rosmarinus
Family: Lamiaceae
Type: Bushy shrub (60–150 cm)
Main growing areas: Morocco, Italy, Portugal, Canary Islands
Harvest: May–November
Parts used: Leaf and flowery stem
Treatment: Steam distillation (essential oil)
Olfactive family: Aromatic
Facets: Fresh, herbaceous, rustic, camphorous, woody, animal

Bourbon vanilla
Madagascar

'This spice with a hundred faces is smoky and animalistic – but, above all, it has a warmth that is ideal for the skin.'

The Aztecs revered vanilla, calling it the 'black flower', in reference to the colour of its pods. They primarily used it to alleviate the bitterness of *xocoatl*, a drink based on cocoa. The conquistador Hernán Cortés was so enraptured by its striking smell that he took some specimens back to Spain with him in 1521 – and the Spaniards christened this new discovery *vainilla*, from *vaina*, the Spanish word for 'pod'.

Around 1880, *Vanilla planifolia* was introduced by planters from the island of Réunion to Madagascar, whose climate closely resembles that of the plant's native Mexico. This 'Bourbon' vanilla is still grown today in northeast Madagascar, on what has come to be known as the Vanilla Coast. This area enjoys an ideal equatorial climate, nourished by the trade winds and abundant rains of the Indian Ocean.

The delicious vanilla pod originates on a creeper from the orchid family, up to 20 metres long, which attaches itself to the bark of other trees with small hooks in order to climb towards the sunlight. Flowers grow in clusters under the leaves, and open only for a few hours. They are pollinated manually by 'matchmakers', who use a long thorn from a lemon tree to bring the pistils into contact with pollen. (The *Melipona* bee that was responsible for pollination in Mexico does not live in Madagascar, so human substitutes have to perform this highly delicate and painstaking operation.)

The fertilization of the flower gives rise to a patient wait of around nine months before the appearance of a pod, which is picked while it is still green and odourless. There follows a long process to obtain the incomparable essence of vanilla. In order to make sure that the pods ripen at exactly the right moment, they are briefly heated in water at 65 °C (bringing the ripening cycle to an abrupt halt). They are then put under blankets for twenty-four hours to remove any humidity. Over the course of the following two weeks, they are dried on large cotton sheets for a few hours every day, in the open air – in the sun at first, but then in the shade. Over this period, the pod acquires both its well-known brown colour and the quintessence of its aromas, as a result of the production of vanillin via the action of enzymes. (Vanillin, which has been synthesized in laboratories for many years, accounts for the most mouth-watering of vanilla's facets, reminiscent of vanilla sugar.) These meticulous, time-consuming manipulations explain the high price of Bourbon vanilla, which experts liken to *grand cru* wines. The offspring of a beautiful orchid, this tropical treasure has doubtless inherited not only its sweetness and nobility but also other equally attractive accents – woody and amber, spicy and smooth – that could legitimately find a place in almost any perfume. Moreover, the extreme sensuality of 'true' vanilla's contact with the skin enhances its appeal still further.

Infusion – maceration in alcohol for several weeks – was the first extraction technique to successfully capture the smell of vanilla, but it is rarely employed today because of its exorbitant cost. Instead, crushed pods are used to obtain a vanilla resinoid with leathery, woody tones via extraction with volatile solvents. The secret of the Aztecs has been revealed to modern perfumers, who continue to venerate the beauty of this most delicious of all orchids.

Genus and species: Vanilla planifolia
Family: Orchidaceae
Type: Low-branching climber (up to 15 m)
Main growing areas: Madagascar, Réunion
Harvest: July–September
Part used: Dried pods
Treatment: Extraction with volatile solvents (absolute), supercritical CO_2 extract, infusion in alcohol
Olfactive family: Spicy
Facets: Woody, tobacco, balsamic, leathery, spicy, creamy

Ylang-ylang
Madagascar

'This languorous, sensual flower evokes, more than any other, the mystery of the faraway, a journey to distant climes.'

This beautiful native of the Moluccas (eastern Indonesia) travelled widely in the southern hemisphere before finding its promised land in Madagascar – more specifically, in Nosy Be, known as the 'island of perfumes'. The effect is exhilarating when the sunny, spicy aroma of ylang-ylang mingles in the air with the scents of coffee and pepper. The small flower with its six golden petals, barely visible among the large leaves, fully lives up to its reputation as the 'flower of flowers' – a nickname apparently picked up in the Philippines – as soon as the nose catches its complex, intoxicating perfume, which permeates its surroundings at sunset.

In its original habitat, with no outside intervention, *Cananga odorata* can reach heights of up to 30 metres and live for fifty years, remaining productive all the while. When cultivated, however, it is regularly pruned to prevent its sinuous branches from soaring into the skies, like the outstretched hands of a wizard. This drastic treatment facilitates the harvest, without affecting the tree's generous output: it blooms several times a year, and the more its flowers are picked, the more it produces: 10–15 kilos a year, on average, for an adult tree.

At dawn, pickers collect the small, extremely fine flowers, choosing only the mature ones, recognized by their deep yellow colouring and a small red mark in the centre of the petal. The flowers are tossed into baskets woven from black vines and then processed immediately in nearby workshops, in order to retain their olfactory qualities as much as possible.

The fresh flowers are distilled in water for up to twenty-four hours, in a still heated over a wood fire. Ylang-ylang is exceptional in that its distillation process is divided into several time periods, resulting in a number of different extracts, each with its own sharply defined olfactory profile. During the first few hours, the flowers release their most precious elixir. After two hours, ylang-ylang 'extra', with its perfectly balanced floral and fruity tones, is ready to be removed; an hour later, ylang-ylang I is available; then, after six hours in all, ylang-ylang II is complete; and, finally, at the end of the process, there is ylang-ylang III, which has a markedly smokier profile. It is also possible to perform the distillation in one go, without these fractioned stages, resulting in a complete essential oil. The heavenly ylang-ylang is truly one of the most bountiful of all flowers. It stands apart for its phenomenal abundance of essential oil: 50 kilos of its flowers are enough to produce a kilo of essential oil, while, by contrast, no less than 4,000 kilos of rose petals are required to provide the same amount of essence.

It is hardly surprising that the essential oil of ylang-ylang is highly prized in aromatherapy for its soothing and de-stressing properties. Its soft, sensual fragrance overflows with warmth, sunshine and gratitude, enfolding and reassuring its recipients before sweeping them away to places where it is easy to live and let live.

Genus and species: Cananga odorata
Family: Annonaceae
Type: Evergreen tree (30 m)
Main growing areas: Madagascar, Comoros, Mayotte
Florescence: All year round
Part used: Flower
Treatment: Fractional distillation (essential oil)
Olfactive family: Flowery
Facets: Floral, spicy, sunny, soft, fruity (banana)

The Americas

Peru balsam

El Salvador

'This esoteric, metaphysical, infinitely tender material reconnects us with a lost world, as if casting a spell.'

According to legend, Princess Naba, daughter of the king of the Nahua, was in love with a prince of the Maya, an enemy people. War broke out and the prince's army was crushed. As he lay dying on the battlefield, Naba ran to him. Her father, horrified by the sight of her kissing her beloved goodbye just as he was taking his last breath, drew his sword and killed her. The soil was bathed in the blood of the unlucky maiden. On the very spot where her body had lain, huge trees grew up in memory of her sacrifice, and their resin, blessed with the power to heal wounds, took on the colour of spilled blood. Peru balsam, an aromatic gum from an extraordinary tree, was born.

This legendary tree originates from the Cordillera del Bálsamo, a heavily forested mountain range in the northeastern region of El Salvador. *Myroxylon balsamum* var. *pereirae*, better known as 'Peru balsam', has actually never grown in Peru, contrary to what its popular name suggests. The explanation is simple: during the period of the Spanish occupation, the balsam was taken to the port of Callao, close to the Peruvian capital of Lima, from where it was shipped back to Europe. From then on, this balsam of El Salvador was unjustly saddled with a Peruvian origin.

The resin collection is an arduous process. With the help of a rope, barefoot *balsameros* climb up the tree, where they remove pieces of bark and make incisions halfway up the trunk, allowing the precious nectar to run out. This exudation is encouraged by burning the trunk for ten minutes with flaming twigs, right up to the level of the cuts. The incised area is then covered with a cotton cloth, like a bandage protecting a wound. Up to fifteen incisions of this type are made on various parts of the trunk. Three weeks later, both the cloths and the bark are swollen with the beautiful brownish-red, syrupy resin. The broken bark is plunged into boiling water to recover the first portion of the balsam, and the soaked cloths are boiled in water for an hour to obtain the second. These two portions are then mixed together.

A harvested tree is left untouched the following year in order to preserve its vital energy, and as a result some centuries-old specimens continue to provide resin over and over again.

Apart from its medicinal properties, recognized by indigenous Americans for centuries – the Maya used it to heal wounds – the resin also boasts an exquisite smell. The balsam projects much of its warm, vanilla richness almost as soon as it leaves the tree, but an enormous complexity still remains to be explored – a mixture of honey and candied fruit, of benzoin and caramel, of vanilla and cinnamon. There is plenty of warmth, but even more sweetness. Whether its treatment results in an essential oil with powdery, cinnamon inflections (via vacuum distillation) or a resinoid with more balsamic, amber notes (via extraction with volatile solvents), Peru balsam has a capacity to soothe and to provide a sense of security by connecting us with the memory of lost civilizations.

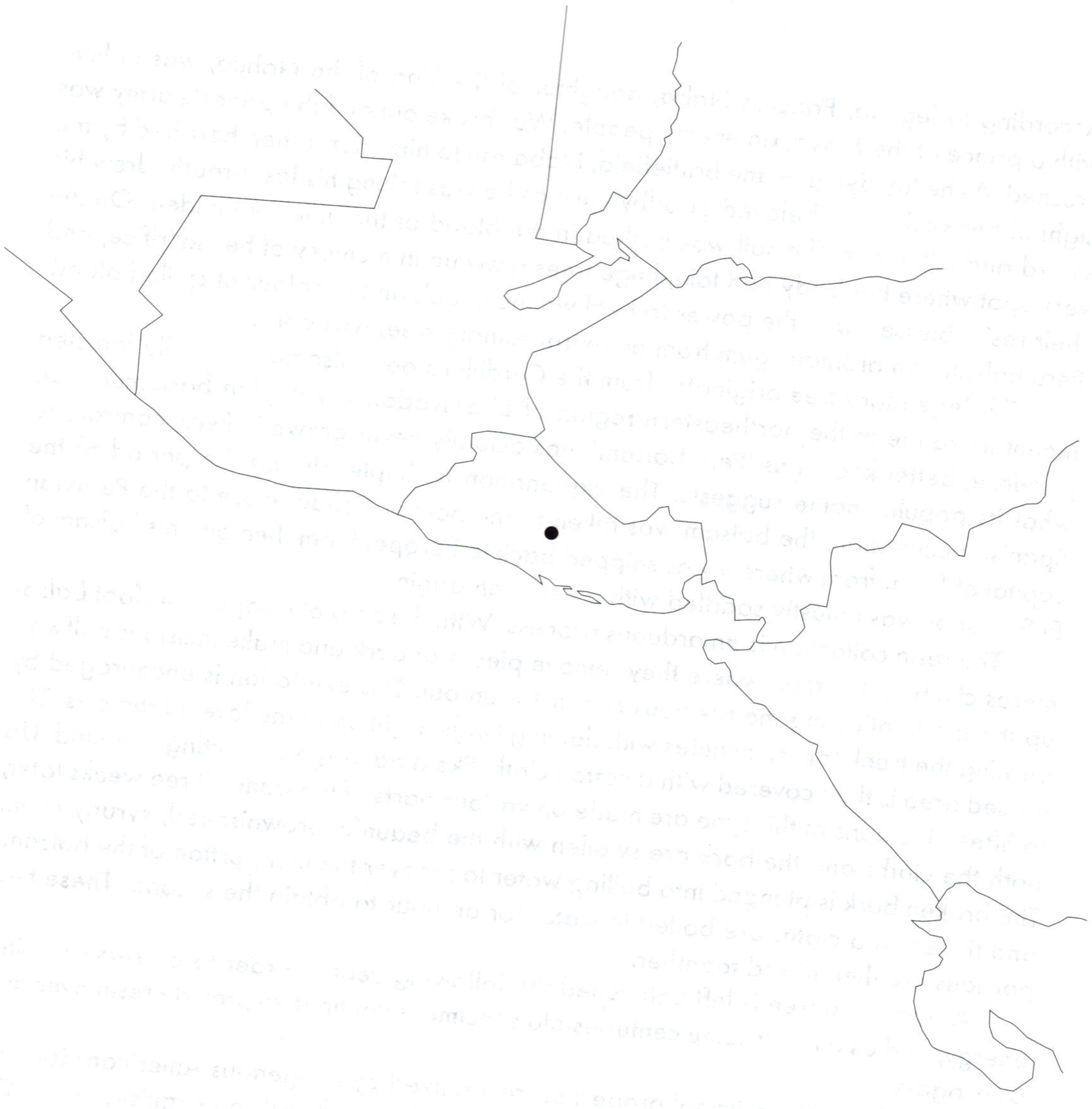

Genus and species: *Myroxylon balsamum* var. *pereirae*
Family: Fabaceae
Type: Monumental tree (25 m)
Main growing area: El Salvador
Harvest: December–May (dry season)
Part used: Resin derived from the trunk
Treatment: Steam distillation (essential oil) and extraction with volatile solvents (resinoid)
Olfactive family: Amber
Facets: Balsamic, amber, sweet, spicy, vanilla, gourmand

Guaiac wood

Paraguay

'This "wood of life" with therapeutic properties exudes a soft smell of great complexity, with a few mysterious notes that recall rose and violet.'

Bulnesia sarmientoi is a tree that largely grows in Argentina, and its wood – known as guaiac or, more popularly, *palo santo* ('holy stick') – was used by Inca shamans as a purifying incense to drive away evil spirits and facilitate communication with the gods. It particularly thrives in the extremely dry subtropical climate of the dense forests of Gran Chaco, on the borders of Argentina and Paraguay – this exceptional biotope is the second most forested region in the world and harbours an extraordinary range of endemic species.

This tree has never been domesticated, but it grows in groups in the wild, on seams of salt that are clearly visible in the ground. It is easy to spot, with its smooth, thick bark, butterfly-shaped leaves and bunches of star-shaped flowers, which are initially blue but later turn pure white.

Traditionally, only dead branches retrieved from the ground after three or four years of decomposition were used, to allow the wood to develop its distinctive essence. Nowadays, the trunk is cut about 20 centimetres from the ground, thereby allowing the tree to fully recover. Once the bark has been stripped off, the fragrant wood is shaved or chopped into thin chips. The wood is almost completely rot-proof, and can be stored for many months without losing its precious olfactory qualities. Its incomparable density – its essence is unusual in that it sinks in water – makes guaiac one of the most resistant woods in the world. According to the Muisca people of Colombia, there were no earthquakes in prehistoric times because the world was supported by four guaiac trees and rendered unshakeable on account of their unwavering strength. The brownish-green guaiac wood has long been popular with carpenters for items where durability is vital: boat propellers, axles, pulleys, balls used in bowling – and judges' gavels. According to legend, it was also once used to make wooden legs for pirates.

Bulnesia sarmientoi has been overexploited, however, probably as a consequence of its reputation and acknowledged therapeutic properties, and it has been added to Appendix II of CITES (Convention on International Trade in Endangered Species of Wild Fauna and Flora). In order to guarantee the survival of the species, the logging of guaiac trees is strictly regulated.

The distillation of chips and shavings from this tree over a period of around thirty hours results in a powerful, compact and solid essence, golden yellow in colour, with woody, slightly balsamic, sandalwoody touches that give rise to a fairly marked rose note. The nose also perceives almost immediately the smell of fire and burnt wood, like a distant echo of the fumigation rituals in which guaiac was traditionally used.

Genus and species: Bulnesia sarmientoi
Family: Zygophyllaceae
Type: Evergreen tree (10–15 m)
Main growing areas: Paraguay, Argentina
Harvest: February–November
Part used: Wood chips or shavings
Treatment: Hydrodistillation of the bark, branches and water (essential oil)
Olfactive family: Woody
Facets: Balsamic, smoky, warm, spicy, leathery

Cardamom
Guatemala

‘Cardamom’s tonifying aromatic endows a perfume with energy, evoking the impression of windows flung wide open.’

Cardamom came to perfumery via cooking. Asian and Arab cuisines revel in it: cardamom is a key ingredient in both Indian masala and North African ras el hanout (a spice mixture available in souks that is liberally applied to tagines and couscous). Cardamom is also an integral part of Indian masala chai and the 'Turkish' coffee that is so popular in Lebanon. It was only after these faraway culinary traditions made a lasting impact in Europe that perfumers began to look to cardamom for inspiration.

For many years, the tea industry monopolized cardamom seeds but discarded the pods, and it was only this by-product that was available to perfumery for any extraction worthy of the name. The resulting essential oil proved extremely shoddy, however, as it incorporated some unpleasant, greasy inflections. Over time, this problem was virtually eliminated through improvements in both the selection of cardamom varieties and the extraction techniques used, resulting in a far more attractive essential oil. These setbacks explain why this aromatic Indian spice made a relatively late appearance in the perfumer's repertoire. Although it formed part of the famous Prunol 960 base produced by De Laire in the 1940s, cardamom only became really familiar to a wider public in the 1990s, when it began to feature on the labels of perfume flacons to the same extent as pepper and saffron.

Elettaria cardamomum (green cardamom; black or brown cardamom is highly prized in cooking but serves no purpose for perfumers) belongs to the same botanical family as ginger and turmeric. It originally came from India – more precisely, the Malabar coast in the southwest of the country – and was introduced into Guatemala a hundred years ago. The plant slightly resembles a reed, because of its extremely tall stems, and in summer white flowers stained with red flourish at its base. The treasure coveted by perfumers is to be found elsewhere, however, in the heart of the fruit. This capsule, similar to a green olive, contains fifteen to twenty small brown pointed seeds that are highly aromatic. The fruits are picked by hand in dry weather and left to dry on mats in the sun for five days, to ensure that they ripen fully (it is crucial to avoid drying them too quickly, as this could cause the fruits to burst). The essential oil of green cardamom is obtained via steam distillation from whole or previously crushed seeds. Another alternative is a supercritical CO_2 extract, which reflects the smell of the seed more closely: it is very lemony and less camphorous than the essence, with a slight 'mango' fruitiness.

Cardamom is classified as a fresh spice and has a relatively volatile scent. Its complexity can sometimes mislead the nose, however. The first impression can be a sparkly upsurge derived from a facet close to rosemary, but once that has evaporated warmer inflections appear, reminiscent of pepper, ginger and clove. This is the surprising nature of a prodigious raw material with multiple lines of seduction that is constantly giving rise to travelled, cross-cultural perfumes.

Genus and species: Elettaria cardamomum
Family: Zingiberaceae
Type: Herbaceous plant (60 cm)
Main growing areas: Guatemala, India (Travancore, Malabar and Canara), Sri Lanka
Harvest: December–February
Part used: Seed
Treatment: Steam distillation (essence) and supercritical CO_2 extraction
Olfactive family: Spicy
Facets: Green, aniseed, camphorous, zesty, aromatic, aldehyde

Red cedar
USA

'This wood is little known to the general public, but it is overflowing with personality and its clear but soft woody note reminds me of a crust of bread.'

In the world of perfumery, the word 'cedar' does not mean very much in itself, since there are several varieties of this tree that bear little resemblance to each other in olfactory terms. While the Atlas cedar exudes leathery animal smells that recall jewelry boxes in the souks of Marrakech, the eastern red cedar, or Virginia cedar, presents a soft, smooth, woody fragrance that is more difficult to pin down. We are all familiar with it, however, even unknowingly: we have only to go back to our schooldays and remember the not altogether unpleasant taste left in our mouth by the pencil we used to chew.

This tree originated in the American states of Virginia and Texas and now grows on poor mountain soils in an area extending from the Saint Lawrence River down to Florida, in the eastern part of the USA. The eastern red cedar can be distinguished by its slender canopy and its green foliage with bluish glints. Its soft pinkish-red wood gives off a strong, delicious smell of resin that repels moths – which is why pieces of cedar are hung in wardrobes. Its resistance to rot and its unrivalled longevity – the tree can live for up to 450 years, and its wood is equally durable – make it ideally suited to the construction of houses, furniture, fences, chests, shoe trees, and more.

The specimens of *Juniperus virginiana* specially selected for the perfume industry tend to be young, at around sixty years. Unlike the common juniper, it is not the tree's berries or twigs that are distilled, but the heart of its trunk; its roots are left in the ground. The wood is cut into logs before being reduced to chips or sawdust, and these by-products are then subjected to six hours of steam distillation. The resulting essence develops a characteristic smell of clean, dry wood. Fractional distillation produces a rectified essential oil, free of its 'head' – that very pronounced and identifiable pencil-lead element.

This essence is categorized as 'vertical' as it can run through an entire composition, from the head to the base, and linger within a perfume until it has completely evaporated off the skin. Accordingly, it is considered as a support in a composition, capable of lending structure. Another asset for perfumers is its liveliness, which imbues a scent with flair and a certain edginess. The essential oil possesses a similar ability to bring together other woody notes in the perfumer's palette, such as sandalwood, vetiver and patchouli. Still, eastern red cedar rarely crops up on the labels of perfume flacons, suggesting that it is not as highly appreciated by either perfumers or consumers as other woods. One explanation could be the lack of any exoticism associated with it; or perhaps it is passed over on account of its plain-spoken directness.

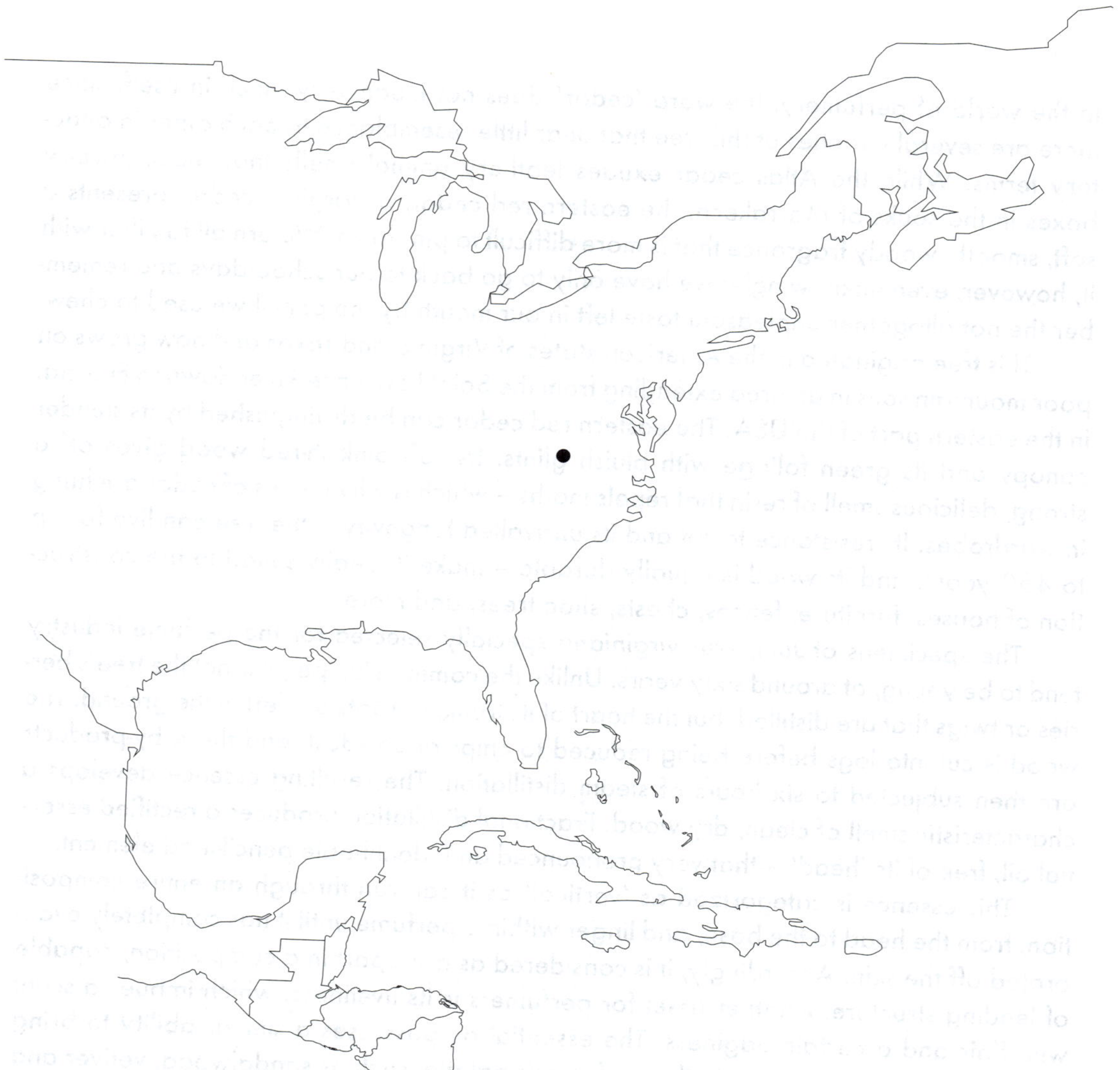

Genus and species: Juniperus virginiana
Family: Cupressaceae
Type: Evergreen conifer (4–20 m)
Main growing area: USA (Virginia and North Carolina)
Harvest: All year round
Part used: Wood chips
Treatment: Steam distillation (essence)
Olfactive family: Woody
Facets: Woody, dry, pencil lead, resinous

Tonka bean
Venezuela

'The aroma of this magical little seed is spellbinding. It is warm and sensual, providing complex flavours without straying into any overly sweet gourmand notes.'

The tonka bean is considered a lucky charm by some indigenous peoples of Amazonia. In parts of Guyana, for example, anybody swearing an oath is traditionally expected to hold a tonka bean in their left hand – along with a dead snake in their right one. Meanwhile, the Panare and Piaroa peoples attribute almost magical medicinal properties to this little brown seed, which they call *sarrapia*. It comes from the Brazilian teak (*Dipteryx odorata*), a tropical tree that has never been cultivated by humans and is classified as 'emergent' – it grows higher than its neighbours in order to guarantee its share of sunlight. It grows spontaneously in dense, humid forests, attaining a height of up to 30 metres. Its oval fruits, reminiscent of a mango, ripen in winter and fall naturally to the ground, where they are gathered by *sarrapieros* – 'bean-pickers' – between February and April. An adult tree can produce around 15 kilos of fruit per year, depending on climatic conditions. The locals have to take security measures to protect this precious treasure from poachers keen to exploit its market value.

Immediately after the harvest, a long process is set in motion. The husks are dried and cut open with a knife, or smashed with a carved stone or hammer, to release the coveted seeds – just one per fruit. These are then left to dry for several months, first in the shade and then in the sun. These stages are crucial to the progressive formation of small, frosty white crystals on the wrinkly surface of the black seeds, indicating the development of coumarin, the fragrant component largely responsible for the passion the tonka bean generates. (The presence of coumarin – at a concentration of 1 to 1.3 per cent in the highest-quality seeds – gives ground beans their powerful aromas of vanilla, chocolate, cut hay and almond.) After a subsequent desiccation the beans are sorted, in a ritual as popular as it is important: any that are too black, too swollen or too soft are discarded, as they will not be able to deliver their full olfactory richness.

The tonka bean was initially packed into small sachets that, like those containing lavender, were used to aromatize cupboards and clean linen; it also added its soft, warm scent to snuff. It was not until the early twentieth century that perfumery discovered the enormous generosity of its elegant, enticing aroma of honey, praline, almond and tobacco, which unobtrusively brings to mind patisserie without overdoing the unpleasant sugary effect. Around a decade ago, the tonka bean also began to make its way into food, as cooks have succumbed, in their turn, to its warm, vanilla taste (even more complex than that of vanilla itself). It is not surprising that this sophisticated sweetmeat has become so popular among chefs, who grate it very finely over crumbles, mashed potatoes and scallops to infuse them with its smooth, full accents. In short, the tonka bean is a gift to the senses – all the senses.

Genus and species: Dipteryx odorata
Family: Fabaceae
Type: Large tropical tree (25–30 m)
Main growing areas: Venezuela, Guyana, Brazil
Harvest: February–April
Part used: Bean
Treatment: Extraction with volatile solvents (resinoid then absolute)
Olfactive family: Powdery
Facets: Almond, warm, tobacco, soft, grilled, hay

Ambrette seed
Ecuador

'The magic of a seed that is both rustic and luxurious, with velvety, silky notes that immediately brighten up any perfume.'

Ambrette seeds (harvested from the abelmosk bush), which can justifiably be called a treat for adults, are habitually used to aromatize both Egyptian coffee and traditional liqueurs such as Benedictine and Poire Williams. It is hard to tell, however, exactly which plant lurks behind that clean smell redolent of pear, even after sniffing a fistful of its seeds.

Abelmosk is a herbaceous tropical plant that originated in India and now flourishes in Ecuador, to the north of Guayaquil. It belongs to the same family as the mallow but is often mistaken for hibiscus (hence its former botanical name of *Hibiscus abelmoschus*), on account of its large flowers, which range in colour from watermelon-pink to creamy white and last only for a single day. Once they have opened, they wilt and make way for strange, hairy pods containing plump, curled seeds that look like small striated snails: these are the source of that coveted fragrance. Predators are not slow to search them out – in fact, some parrots love these seeds so much that farmers have to hire guards armed with blank-firing rifles during harvest season to scare them off. Once gathered, the treasured grains face a further danger, in the form of a parasite. As a precautionary measure, the harvest is transported immediately to the Ecuadorian capital of Quito, at an altitude of almost 3,000 metres, where the potential aggressor has no chance of surviving.

The ambrette seed has to be distilled with steam under pressure for a whole forty-eight hours, as it is not particularly generous with its essence. The resulting essential oil is called ambrette butter, like iris butter, on account of its paste-like consistency. There then follows a second transformation process, which involves eliminating all the fatty acids with an alcohol wash to obtain the end product incorrectly referred to as 'ambrette absolute'. It is an upside-down world where an essence is called an 'absolute' to give it an extra touch of class, even when it oozes class in the first place. It is not for nothing that this essence is one of the most expensive of all the raw materials used in perfumery – notwithstanding the fact that price is not always an accurate measure of quality.

Ambrette is highly seductive and capable of endowing a perfume with something unusual and delicate: texture. In other words, not only can its seed be smelled, it can practically be seen and touched as well, to such an extent that it seems real, with its frugally woody, powdery-iris, almost dusty inflections that conjure up images of South America. Then comes a soft, cottony facet so unexpected that it demands another sniff. Ambrette is a raw material of exceptional tenacity that can be described as 'vegetal musk', thanks to the natural presence of ambrettolide, a highly musky, powdery molecule. Finally, ambrette's vibrancy and elegance would not be quite the same without its delicious pear facet, which enhances its character and provides a gateway for newcomers to its wondrous scent.

Genus and species: Abelmoschus moschatus
Family: Malvaceae
Type: Herbaceous tropical plant (c. 1.3 m)
Main growing areas: Ecuador, Peru
Harvest: May–August
Part used: Seed
Treatment: Steam distillation or supercritical CO_2 extract (essential oil, known as absolute)
Olfactive family: Musky
Facets: Woody, iris, earthy, fruity (pear), liquorous

Key lime
Mexico

'This pleasing, effervescent note is as sizzling as it is joyful, and it contributes brilliance and a touch of exoticism to a perfume.'

The summery accents of the key lime conjure up a picture-postcard paradise where exhilarating cocktails such as caipirinha, mojito and ti' punch are served. In similarly festive mode, in Mexico key limes are cut into quarters and slipped into the neck of a beer bottle before drinking.

The hesperidium (citrus) family offers perfumers a seemingly unlimited supply of aromatic wonders, ranging from bergamot to grapefruit, from mandarin to citron. Some of these are stars, while others take on a supporting role, but even the lesser-known performers make a crucial contribution to the success of a composition. Although *Citrus* x *aurantifolia*, or key lime, closely resembles a small green lemon, it must not be mistaken for the similar fruit of *Citrus* x *latifolia*, the Persian lime, commonly known simply as lime. The fact that key lime has acquired an array of nicknames ('West Indian lime', 'bartender's lime', 'Mexican lime') only adds to the confusion. Physically, it looks like a cross between different citrus fruits – an orange with the skin of a bergamot. Its high concentration of vitamin C led to the consumption of key lime juice on British ships in the nineteenth century to combat scurvy – to such an extent that British sailors became known in the United States as 'limeys'.

Little known in Europe, this citrus fruit with its strong character probably originated in the mountainous regions of Meghalaya and Nagaland in northeastern India, but it has now been naturalized in Mexico, Peru, Argentina and most of the Caribbean. It is sensitive to the cold and cannot tolerate temperatures below zero. In tropical climates, the key lime tree, with its dark green foliage and upright, lemon tree-like silhouette, is covered with white, star-shaped flowers all year round, although they are particularly abundant in March and April. The fruits appear six months after the florescence, in the form of small, roundish lemons, no more than five centimetres in diameter, with a 'nipple' on the end. Their thin, smooth rind and juicy, acidic, seedless pulp are widely used in Caribbean and Asian cooking, for example in dishes such as ceviche and the Creole jambalaya.

Key limes are picked by hand and then taken to be sorted and washed. In contrast to all the other citrus essences, that of *Citrus* x *aurantifolia* is obtained through crushing and distilling the whole fruit. This is first put into a press, and the resulting juice is distilled with steam. Connoisseurs have found that the scent of key lime essential oil is very faithful to that of the ripe fruit.

Although much of the production of acidic key lime finds its way into fizzy drinks, perfumers also appreciate its resemblance to an acid drop sweet and its capacity to refresh a composition by introducing vivacity and joyfulness, 'pep' and tonicity. Key lime was long confined to functional perfumery and detergents, but it has finally found the place it deserves in a nobler arena.

Genus and species: Citrus x aurantifolia
Family: Rutaceae
Type: Thorny shrub (5 m)
Main growing areas: Mexico, Peru, Argentina, South Africa, the Caribbean
Harvest: March–April
Part used: Fruit
Treatment: Steam distillation (essential oil)
Olfactive family: Hesperidia
Facets: Fresh, lemony, green, zesty, sizzling, tropical

Peppermint
USA

'Simple mint can play a very "clean" role or enliven a perfume with a natural, radiant and refreshing breath of fresh air.'

Mentha is derived from the name of the nymph Minthe in Greek mythology, with whom Hades, god of the underworld, fell obsessively in love. Persephone, Hades' wife, was so jealous that she frenziedly trampled her rival before turning her into an aromatic plant: mint. Hades was overcome with pity and gave Minthe a heady smell in order to preserve her memory forever. Ovid, who told this story in his *Metamorphoses*, was right: the pervasive and fresh herbaceous aroma is indeed entrancing.

Of the many varieties of mint in existence – spearmint, pennyroyal, and so on – peppermint is certainly the most easily recognizable, on account of its lively, intense 'chewing gum' aroma. *Mentha x piperita*, as it was named in the seventeenth century by the English botanist John Ray, was the result of a natural hybrid between green mint, *Mentha spicata*, and water mint, *Mentha aquatica*. This hybrid has gone on to be probably the most popular fragrant plant in the herbal repertoire, and it has provided the basis, in its turn, for eau de Cologne mint, as well as contributing to *essence d'agrumes* perfume and chocolate mints such as Britain's famous After Eights. According to botanists, mint is probably the easiest of all plants to hybridize: it is sufficient to plant two different mint plants side by side and a new variety will emerge. There is no mistaking peppermint when you see it growing, with its serrated, oval, deep green leaves ending in a point. Its purple flowers cluster together around the top of the stem, but these are irrelevant to perfumers, who are interested only in the leaves – even though peppermint is unstintingly generous, as every part of it is strongly aromatic.

Several terroirs compete for the honour of growing the best peppermint, but one of the most fragrant crops is to be found along the 45th parallel in the Yakima Valley, in Washington State, USA. This region, like Montana and Idaho, offers climatic conditions suited to peppermint, in particular constant, heavy rainfall throughout the year, which increases the plant's concentration of menthol. It is this component that provokes that familiar soothing and refreshing sensation, but it is abundant only before noon and drops off considerably later on – so mint is harvested exclusively in the morning, before it is in bloom, so that the active substances are more concentrated in the leaf. The plant is cut down at ground level and then left to dry in the open for a few days, like hay. Once it has been sufficiently exposed to the air, it is stored in a dry place. The essential oil of peppermint is obtained by distilling the leaves, which are separated from the stems just beforehand. The result has a direct, open quality, an untrammelled naturalness coupled with a tingling freshness. Spearmint is also used in perfumery, but it is softer and more rounded. Between the two, they ensure that Minthe is still flourishing today.

Genus and species: Mentha x piperita
Family: Lamiaceae
Type: Herbaceous perennial plant (50 cm)
Main growing areas: USA, France, Italy, Japan, Morocco
Harvest: June–July
Part used: Leaves
Treatment: Steam distillation (essential oil)
Olfactive family: Aromatic
Facets: Fresh, herbaceous, green, icy, camphorous, smoky

Vetiver
Haiti

'This root, with its wild beauty and fiery temperament, imbues a perfume with warmth and sensuality, turning it into something comforting.'

Vetiver plunges deep into the earth, tirelessly soaking up its most vital resources. Its roots, which can grow up to two metres in length, are thin, hairy and densely intertwined, thus serving to stabilize slopes and reduce erosion. And it is in these roots that vetiver's unforgettable raw and organic smell is hidden.

Vetiver originally came from southern India but is now cultivated in various tropical areas. Its aroma is particularly enlivened by volcanic soils, and two of the world's most active volcanic zones – Haiti and Java – provide perfect breeding grounds for the plant, thanks to their exceptionally rich soils replete with mineral salts.

Vetiver is associated with lushness: it presents a note considered 'woody' – like sandalwood and cedar – which is complemented by green, vegetal inflections redolent of crumpled leaves and sap. Clearly the soil in which this lively plant lives will exert a huge influence on the construction of its olfactory profile. Perfumers generally consider the Haitian variety the ultimate expression of vetiver, which displays more foresty, iris-root effects and far less smokiness than its counterpart from Indonesia. There is an almost unanimous consensus among experts that it presents the most harmonious balance between cool and warm tonalities, although everything ultimately depends on the inflections that a perfumer wishes to exploit – at one extreme, there is a sweet naturalness; at the other, a kind of archaic brutality.

A demanding plant that requires great patience, vetiver needs to remain in the ground for twelve months before its roots reach their olfactory maturity. Subsequently, the first step is to cut off the visible part of the plant (its grassy tufts) at ground level. The roots are then removed from the ground, with substantial help from a pickaxe: it is not for nothing that a vetiver harvest is also known as a 'dig', because the soil has to be constantly dug and turned over, by increasingly aching arms. The roots are painstakingly cleaned of any earth clinging to them and the remaining vetiver stumps are transplanted, ready for the next harvest. The roots are then washed, cut and left to dry in the sun, before being subjected to a transformation process lasting at least twenty-four hours. Often carried out in poorly equipped facilities, distillation requires real skill and constant vigilance to avoid 'burning' the end product; in the past, it was not uncommon for essences to give off a somewhat unpleasant hint of charred bacon. The thick, resinous elixir is an attractive deep brown in colour, with a smell of humus, undergrowth and smoked wood. Fractional distillation results in a 'heart' of vetiver: this is a noticeably less earthy and fresher, more exuberant version of this entrancing root, complete with delightful touches of grapefruit – fully justifying the Haitians' description of vetiver as a 'miracle plant'.

Genus and species: Chrysopogon zizanioides
Family: Poaceae
Type: Herbaceous plant (1.5–2 m)
Main growing areas: Haiti, Réunion, Indonesia (Java), China
Harvest: January–July
Part used: Root
Treatment: Steam distillation (essential oil)
Olfactive family: Woody
Facets: Woody, earthy, smoked, fruity (grapefruit), peanut

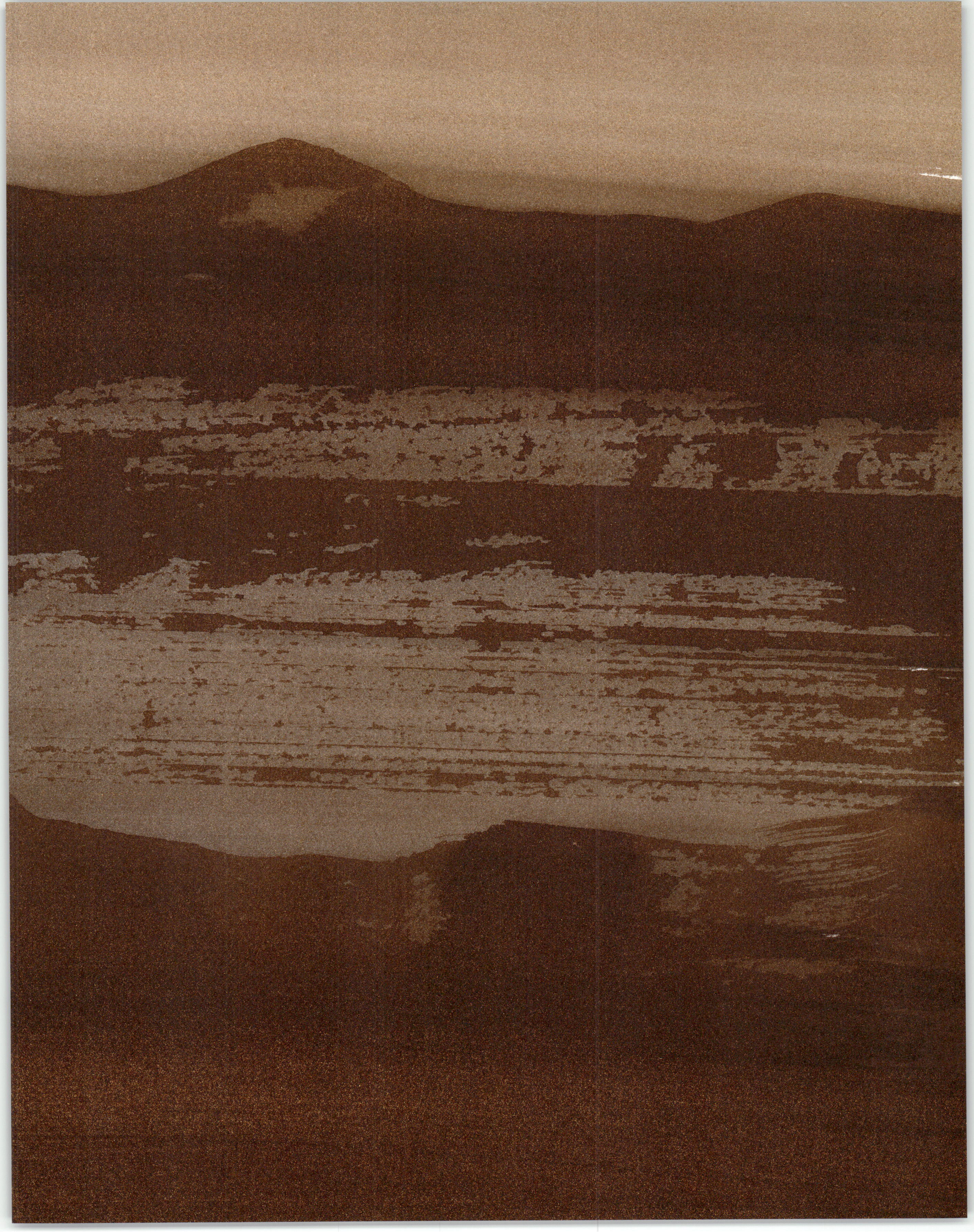

Asia and Oceania

Star anise
China

'This spice with aniseed flavours brings back memories of sweet childhood treats and is probably the sweetest of all the fresh scents.'

The smell of star anise, also called badian, from the Persian *bādiān*, meaning 'anise', 'fennel', invariably evokes Mediterranean aperitifs such as pastis and ouzo, which are packed with sunshine and joie de vivre; but the tree that produces this spice has nothing to do with southern Europe. It originated in southern China, in the province of Jiangxi, where it grows in the mountains at an altitude of 2,000 metres. It has also found a home in the tropical climates of Vietnam, Cambodia, Laos and the Philippines. Its name of star anise can lead to confusion with the green anise produced by *Pimpinella anisum*, a European herbaceous plant.

The tree's pyramidal canopy and splendid bright green foliage are reminiscent of the magnolia, another tree with a Chinese background. The strong, distinctive anise smell comes from the fruit, although the bark provides an exquisite foretaste of the fruit, leading some botanists to refer to it as 'anise wood'. The tree is in no hurry to offer up its essence: it can easily live for a hundred years, starting to bear fruit only in its seventh year of life.

The harvest takes place twice a year, in spring and autumn, and is still undertaken by hand, usually using bamboo ladders to reach the highest branches. The fruit, which is picked before it is completely ripe, is shaped like an eight-pointed star. Once the harvested fruits have been dried in the sun, they take on their familiar brownish-red colour.

A few centuries ago, star anise was so precious that it enabled China's southern provinces to pay off all the taxes imposed on them by the emperor. Star anise was once used in traditional Chinese medicine to refresh the mouth and ease digestion – elderly people would chew the seeds shortly after eating a meal. Even today, its decongestant properties have made it a popular remedy in China for bronchitis and phlegmy coughs. And a molecule extracted from star anise serves as an active substance in the anti-influenza medicine Tamiflu.

Star anise is widely used in Asian cuisine, particularly as a component of the famous 'five spice' mixture (alongside Sichuan pepper, cinnamon, clove and fennel seed), a popular seasoning for fish, meat, dim sum and other dishes. Its light, sweet note conjures up the taste of liquorice – so it is hardly surprising that it whets the appetite of perfumers, those great connoisseurs of flavours. The essential oil of star anise is obtained via distillation of its seeds and bark in a still over a naked flame. The transformation process takes forty-eight or sixty hours, depending on whether the fruits are fresh or dry (the leaves are sometimes distilled as well, but they produce an essential oil of considerably poorer quality). The lively and agreeable freshness of star anise makes it a natural choice for perfumed accords classified as 'fern' – lavender, geranium and coumarin – or as woody and very masculine – on account of its dynamism, as well as its more gustatory and regressive notes of gingerbread.

Genus and species: Illicium verum
Family: Schisandraceae
Type: Evergreen tree (8–15 m)
Main growing areas: China, Vietnam, Cambodia, Laos, Philippines
Harvest: April and October
Part used: Seed and bark
Treatment: Steam distillation (essential oil) and supercritical CO_2 extract
Olfactive family: Aromatic
Facets: Herbaceous, fresh, anise, liquorous, spicy, woody

Benzoin
Laos

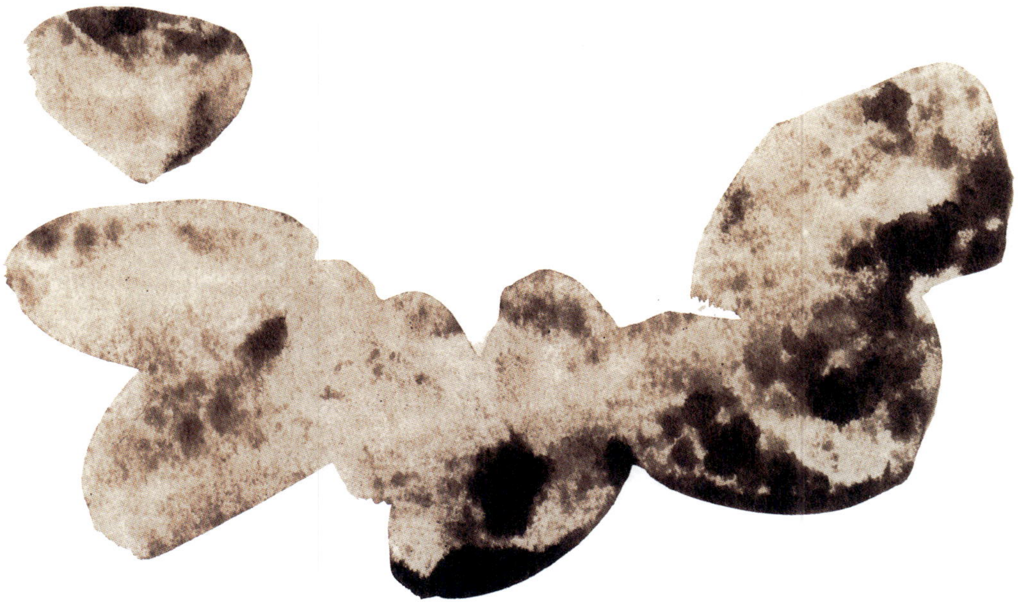

'This irresistible resin with the sweetness of sugared almonds has bestowed all its generosity and goodwill on the perfumery of the twenty-first century.'

It takes plenty of walking to get the chance to see Siam benzoin 'weeping' from a tree. *Styrax tonkinensis*, a large tree resembling a birch, grows wild at altitudes above 1,200 metres in the provinces of Phongsaly and Houaphan in northern Laos, a misty, mountainous region with gigantic forests that is a phenomenal repository of biodiversity.

The tree's precious sap is collected by hand, with the help of a simple knife, as it cannot secrete its extraordinary balsam without first undergoing a trauma that stimulates exudation as a form of defence. Starting at the foot of a tree, a tapper makes a series of horizontal incisions in the bark, 15 centimetres long. He gradually climbs to a height of two metres, making new cuts on the way. Once the trunk has been slashed, the balsamic gum will literally weep from the cuts – although not straight away, starting three to four months after the tree is tapped. The remaining bark acts as a reservoir for the oozing sap. When the tapper climbs the tree once again, in the summer, he can scrape off the beautiful, solidified orange 'tears' that have formed on the surface of the incisions over the previous weeks. Sometimes the tappers encircle a tree before any incisions are made and serenade it with love songs to prepare it for the generous offering of its delicate balsam.

Benzoin resin collection is a precarious business in financial terms, as a tree starts to produce resin only after seven years and the supply is exhausted by the time it reaches eighteen years, so the local farmers combine the work with growing rice on the mountain slopes.

After collection, the sap is then washed and cleaned of any impurities or remains of bark – work that is usually performed by women. The resin tears are then separated according to their size and colour, before being assigned a particular grade (from A to D) that indicates their specific olfactory properties. The highest quality, distinguished by strong notes of vanilla, comes from larger tears. The latter's 'old ivory' colouring immediately distinguishes Siam benzoin, *Styrax tonkinensis*, from Sumatra benzoin, whose tears are reddish, with only a faint smell, and lacks its counterpart's finesse and complexity. The gum of Sumatra benzoin, found in Indonesia, is mainly processed industrially as a flavouring agent. The nobler and more highly prized Siam variety – which, like incense, is associated with a sense of the sacred – has long been a favourite with perfumers. Siam benzoin's smooth, liquorous, honeyed, almond and caramel facets enable it to bring to a composition both strength and sweetness, both richness and comfort. Some tears are sweeter than others!

This resin's delectable smell was popularized in the nineteenth century by Auguste Ponsot, a chemist, and Henri Rivier, a pharmacist, who introduced it into the everyday life of the French people. These two men joined forces to invent Armenia paper, a kind of blotting paper perfumed with benzoin resin that was burned to fill a house with fragrance – echoing the ancient practice of burning incense to ward off evil spirits and sickness.

Genus and species: Styrax tonkinensis
Family: Styracaceae
Type: Large evergreen tree (25–30 m)
Main growing areas: Laos, Vietnam, Malaysia, Indonesia
Harvest: January–February
Part used: Resin
Treatment: Extraction of the resin with alcohol (resinoid and then absolute)
Olfactive family: Amber
Facets: Vanilla, resinous, balsamic, caramel, milky, spicy

Cinnamon
Sri Lanka

'The oldest spice in the world radiates the scent of an exotic, burning light and ends up winning over our hearts.'

Cinnamon is the guest of the dessert, where it exalts the flavours. A pinch of cinnamon is like a touch of amber, insinuating itself into rice pudding, pancakes and crumbles. Perfumers were not slow to succumb to this delicacy and its dreams of escape. Whole 'cinnamon gardens' can be found in the southern part of the island of spices that is Sri Lanka, on a coastal strip dotted with banana and coconut trees. This tree with luminous green leaves belongs to the camphor family. It can reach a height of 10 to 15 metres, but for the sake of convenience it is maintained in its bushy state (around a quarter of its potential size) on commercial plantations. Cinnamon harvesters are known locally as 'peelers' and were traditionally members of the Salagama (or Chaliya) caste, associated with the cultivation of cinnamon since time immemorial.

Twice a year, the young cinnamon branches are cut when the sap starts to rise. The light-brown papyrus-like outer layer is removed before the tree's inner bark is peeled off and cut into pieces of equal length. The best yield is obtained just after the rainy season, as the bark is easier to peel off when it is swollen with water. Dried for several hours in both sun and shade, these fragile shavings curl up into the hollow sticks, or 'quills', that we are so familiar with. By this stage, cinnamon presents its characteristic caramel colour, which provides a foretaste of its aroma. Once the quills resemble miniature rolled-up parchments, they are ready to be distilled.

In terms of smell, Ceylon cinnamon – that is still the accepted designation, even though the island was renamed Sri Lanka in 1972 – has no competition. Its Chinese rival, *Cinnamomum cassia*, is distinguished by its peppery notes, which are far more prominent that any warm notes, but it is noticeably less sweet, as well as more pungent and bitter, and it cannot hold its own against the warm, sophisticated aroma of its relative. In contrast, the delicately velvety Ceylon variety brings a heat and gourmand sweetness to a composition that is difficult to resist.

The essence of cinnamon is its most highly prized presentation, derived from steam distillation of the bark that has been macerated in water. The result is a pale-yellow liquid with an enticing smell and the allure of faraway places; the essence derived from the central section of the trunk is the most sophisticated of all. It is also possible to distil the leaves, petioles and young branches of the cinnamon tree to obtain a spicier essential oil, closer to carnation and clove.

Cinnamon, considered to be an 'olfactive tiger', is strikingly strong and feisty, and it requires a firm hand from any perfumer determined to introduce it into a composition. Experience brings with it the skill of dosing cinnamon sparingly to render it more tender and delicate; in these proportions, it releases its most attractive fruity and vanilla tones.

Genus and species: Cinnamomum verum
Family: Lauraceae
Type: Tropical evergreen tree (10–15 m)
Main growing areas: Sri Lanka, Seychelles, Madagascar
Harvest: April–October, after the rainy season
Part used: Bark
Treatment: Steam distillation (essence) and supercritical CO_2 extraction
Olfactive family: Spicy
Facets: Spicy, balsamic, warm, woody, sugary, vanilla

Clove
Indonesia

'Clove always exudes an enticing, soothing spicy warmth ideally suited to winter, whether it is evoking the mellow flavour of mulled wine, spice bread or toffee apple.'

For many people, cloves evoke the anxiety-inducing smell of a dental surgery, or the old-fashioned ritual of sticking them into an orange to refresh the air in a house during winter. For perfumers, however, clove is the warmest of spices – bold, confident, even impertinent.

A clove (a Middle English term for 'nail') is in fact the bud of a flower. It is harvested when it is still closed, after it has taken on a pretty pink colour. This stage of maturity is brief, and any cloves harvested once they are in bloom (known as 'headless cloves') are of much poorer quality. The buds are picked by hand in two or three successive 'passages', to ensure that they are collected at the optimal moment. They are gathered skilfully and painstakingly in small bunches known as 'claws'. In a coordinated operation, women and children pick those within arm's reach, while men climb bamboo ladders to the upper parts of the tree, where they shake off cloves that are subsequently collected from the ground below.

Once they have all been harvested, the cloves are separated from the claw – that is, from the bunch of stalks – in a drying area. The buds are spread out on reed mats and exposed to sunlight for three or four days, until they take on the dark-brown colour and shape characteristic of the spice in our kitchens. Meanwhile, fingers are crossed for the rains to hold off long enough to prevent any disruption of the drying process.

Historically, Indonesia, and more specifically the idyllic islands of the Moluccas and the Celebes, was the hub of clove cultivation, and even today it is still the world's foremost producer. Later, Zanzibar also became a serious contender, along with Madagascar, which is now number two overall. The clove tree is related to the eucalyptus and is accustomed to tropical climates. It needs heat and humidity, while also preferring low altitudes – it flourishes particularly well close to the sea.

A classical distillation procedure is used to obtain clove essence; it is technically possible to extract the leaves of the clove tree in the same way, but the resulting essential oil is not put on the market. The powerful and pervasive essence of cloves adds depth to a composition, but it has to be handled with great discretion to avoid highlighting its medicinal facets. After the first impressions, a floral subtext evoking carnation can be discerned (eugenol, which comprises 85 per cent of the clove, also acts as an olfactory marker in carnation). Clove's warm notes even hide some particularly sensual amber notes that take the nose by surprise. All in all, clove is a spice endowed with immense richness, an olfactory landscape all its own.

Genus and species: Eugenia caryophyllata
Family: Myrtaceae
Type: Conical evergreen tree (6–12 m)
Main growing areas: Indonesia, Madagascar, Tanzania (Pemba and Zanzibar)
Harvest: April–July (Indonesia); September–December (Madagascar)
Part used: Bud
Treatment: Steam distillation under pressure (essential oil) and extraction with volatile solvents (absolute)
Olfactive family: Spicy
Facets: Warm, floral (carnation), medicinal, woody, camphorous, smoky, balsamic

Galbanum
Iran

'A little forgotten by the modern world, this noble essence with its raw, green scent conceals within its core a lively, aromatic freshness that is truly addictive.'

Galbanum is a legend in itself – and not only because it forms part of the composition of theriac, an antidote to poisons first described by Andromachus, the emperor Nero's doctor. In perfumery, the very word 'galbanum' suggests audacity and creative panache. Embodying the famous 'green' note of the golden age of the perfume industry, it is one of those legendary beauties who inspires dreams even in those who have never beheld her in the flesh. Even the most adventurous perfumer is unlikely to have actually seen the galbanum's yellow umbrella-like flowers, which give off an unpleasant smell that completely belies the olfactory treasure hidden below the soil. The chance to encounter this odd iconoclast would require a long journey across deserts and over mountains: the galbanum grows wild in the desolate expanses of the Alborz Mountains, to the northeast of the Iranian capital, Tehran. The root of this hermaphroditic plant contains an oily gum that gives the plant its Latin name: *Ferula gummosa*.

The gum is harvested at the height of summer by first making incisions in the galbanum's tap roots, although the stem may also be cut. The viscous resin seeps out and is left to thicken and harden. It is then collected, and a further incision is made in the plant in anticipation of the next harvest, fifteen days later.

The galbanum's natural exudate was used by ancient civilizations in ointments and as incense, as well as being a key ingredient for embalming rituals. Perfumers were quick to see the opportunities provided by the deep green notes in galbanum's olfactory character that were rough and forthright but altogether delightful. Its raw, direct and 'committed' note is, however, neither simplistic nor unambiguous, despite its potency. Galbanum gum has some fifty shades of green, from resinous to crunchy (tomato leaf), from its herbaceous facet to its vegetable option (specifically the freshly shelled garden pea).

The use of galbanum in perfume demands perfect control of a composition. The dosage must be exact, to the milligram, otherwise the smell can quickly become unpleasant. Added sparingly, galbanum supplies the fresh and natural boost of a penetrating energy drink that comes into its own in a floral bouquet – where it can also be extremely useful for reconstituting some notes that are impossible to extract directly, such as hyacinth, narcissus and gardenia. If galbanum is applied in excess, however, the rootiness comes to dominate, leading to sourness and an overwhelming bitterness that sucks the life out of all the other surrounding notes and ends up smothering every effort towards creativity. Unfortunately this drawback has quickly returned galbanum to its status as a museum piece, although some declarations of its obsolescence may also be interpreted as admissions of defeat by perfumers incapable of dominating its boldness and holding in check some of its excesses. The fact is that no other raw material expresses so fully the hope of giving nature a new lease of life than the green of galbanum.

Genus and species: Ferula gummosa
Family: Apicaceae
Type: Herbaceous plant with long roots (1 m)
Main growing area: Iran
Harvest: July–September
Part used: Gum resin
Treatment: Steam distillation (essence) and extraction with volatile solvents (resinoid)
Olfactive family: Green
Facets: Green, woody, balsamic, resinous, fresh, earthy, animal, herbaceous

Arabian jasmine
China

'As sunny and fresh as a summer morning, this jasmine joyfully evokes the smell of the pittosporum shrub that blossoms in springtime on the Côte d'Azur.'

There is no danger of Arabian (or sambac) jasmine entering into competition with the *grandiflorum* variety that grows in Grasse, for the simple reason that perfumers do not consider it to be a fully fledged jasmine. It does in fact stand halfway between orange blossom and *grandiflorum*, with its crispy green facets more marked (and less animal) than the latter's. Arabian jasmine first appeared in compositions only in the late 1980s, but it has since become one of the leading flowers in modern perfumery. (The Chinese variety arrived on the scene even later, barely a decade ago.) This plant is a veritable aromatic kaleidoscope, delighting the senses with its seductive freshness coupled with plenty of zip. Every one of its white flowers is imbued with its main characteristics: the sweetness of orange blossom, the fresh greenness of lily of the valley and the sunny temperament of tuberose.

The climbing shrub with plump, opalescent white petals that has acquired the name 'Arabian jasmine' actually comes from Asia, more specifically the Himalayan valleys, and it was traditionally cultivated in Tamil Nadu, in southern India. Its flowers, which are often used to make ceremonial garlands, are picked before the monsoon, from March to September, at dawn, when the ground is still shrouded in mist retaining the coolness of the night. This plant's botanical name, *sambac*, provides irrefutable proof of its geographical origins: it is derived from the Sanskrit word *campaka*, meaning 'heady perfume'.

Arabian jasmine also flourishes on the other side of the Himalayas. The Chinese have a predilection for this type of natural fragrance, including osmanthus and magnolia, for example. In imperial China, the elegant ladies of the court demonstrated their appreciation of both the visual and olfactory appeal of sambac jasmine flowers by wearing them in their hair. These flowers also started being used very early on to perfume green-tea leaves, giving rise to the beverage commonly known as 'jasmine tea'. When brewed at a constant temperature of 37 °C, the liquid becomes infused with the flowers' sweetness, resulting in a small miracle: the combination of the aroma of jasmine flowers and tea leaves makes the drink far more refreshing – a real-world demonstration of the remarkable properties of this flower with its green, sweet facets. And it was through its presence in tea culture that it gradually caught the attention of perfumers and ended up in their repertoire of raw materials. The Chinese heartland for Arabian jasmine is Heng Xian, in the Guangxi region, at the southeastern tip of the country. It has been grown there for over 400 years, and its roots are also used as a sedative throughout China.

Arabian jasmine is prized by perfumers for its distinctive character: it is less heady, but lighter and more virginal than its *grandiflorum* cousin, while the absolute is reminiscent of a flower intoxicated with sugar and pollen, generously offering its nectar in the damp air of summer. At the same time, its exotic feel enables it to evoke the heat and abundance of Asia with the same conviction.

Genus and species: Jasminum sambac
Family: Oleaceae
Type: Evergreen shrub (1–3 m)
Main growing areas: China (Guangxi, Yunnan), India
Harvest: March–September
Part used: Flower
Treatment: Extraction with volatile solvents (absolute)
Olfactive family: Flowery
Facets: Orange, sunny, tropical, herbaceous, green, fruity (banana)

Magnolia
China

'Any perfume is illuminated by the introduction of this fresh and generous flower, which smells of Asia after the rain.'

The magnolia has existed on Earth for millennia. According to ethnobotanists and archaeologists, fossilized plants belonging to the magnolia family date back to 95 million years ago, so it could have been growing alongside the dinosaurs. It is a delightful paradox that one of the most ancient of all flowers now entrances us with a smell that so strikingly expresses the freshness of a new day.

Although it was named in homage to Pierre Magnol (1638–1715), director of the Botanical Gardens in Montpellier, France, who helped to introduce the classification of plants by 'families', the magnolia tree's ancestral roots lie outside Europe. It originally came from China, Japan and Central Asia, as well as the south of the United States. Of the 245 species that have been identified, the *Michelia* x *alba* variety appeals most to perfumers; it can be distinguished by its twelve thin petals that form a star pointing towards the sky.

This pretty white flower is grown in the southern Chinese provinces of Guangxi and Fujian. It is harvested in fine weather, from May to June, and then from September to November. The magnolia tree used to grow to unmanageable heights – there was even a time when men used to put up scaffolding and risk their necks to reach the flowers on the highest branches. Domestication has gradually reduced the tree to a more reasonable size, thereby facilitating the harvest. The flowers are picked in the early evening (or sometimes at the dead of night), when they are at their most fragrant, before being spread out for a few hours in the sun to dry, prior to being processed. Today, part of the crop is still mixed with green-tea leaves to enhance their aroma and add a soft, sweet perfume.

Gardeners are drawn to the magnolia by its harmonious colour tones and dramatic, abundant flowering, but perfumers are more attracted by its incomparable green and floral smell, redolent of springtime. The magnolia, classified in perfumery as a 'white flower', readily infuses other components of a bouquet with its soaring naturalness; yet among its sensual, extrovert peers, such as jasmine and tuberose, it stands out from the crowd. Its distinctive melody imbues a perfume with an ethereal elegance, more typical of a citrus fruit than a flower. Neither pretentious nor stuffy, it can contribute sunny, vanilla notes capable of delighting even the most jaded nose.

It might be assumed that magnolia essential oil, with its floral, watery and lemony touches, has been around since the beginnings of modern perfumery, alongside rose essence and jasmine absolute, but in fact it did not enter the repertoire until the 1990s. Until then, perfumers had had to make do with facsimiles, reconstitutions of the smell of this flower of varying quality and fidelity from China.

Genus and species: Michelia x alba
Family: Magnoliaceae
Type: Large evergreen tree (up to 15 m)
Main growing area: China
Harvest: May–June, then September–November
Part used: Flower
Treatment: Steam distillation (essence)
Olfactive family: Floral
Facets: Fresh, floral, green, citric, sunny

Nutmeg
Indonesia

'I love nutmeg's incredible complexity, which takes it from hot to cold, from woody bitterness to delicate sweetness in an instant.'

This little odiferous treasure was once the most expensive raw material on Earth, avidly coveted in the sixteenth and seventeenth centuries by the Portuguese, the Dutch and the French. Nutmeg originated in the Indonesian archipelago of the Moluccas, specifically the Banda Islands – volcanic fragments scattered on the Pacific Ocean. This tiny paradise, basking in the tropical sun and almost forgotten by the world, is home to the green, immutable nutmeg tree.

A nutmeg tree's fruits cannot be harvested until seven years after it is planted, as it takes its time to mature; after that, however, they can be harvested three times a year, in spring, summer and winter. According to local custom in the areas where nutmeg is cultivated (Indonesia, Sri Lanka and Grenada), the fruit is picked from the branches with a bamboo pole or simply collected from the foot of the tree. At first sight, it resembles a nectarine or a yellowish apricot, with a fleshy part that splits open when ripe to reveal a brown nut surrounded by a rubbery red membrane resembling a hairnet (this is the spice known as 'mace'). Simply shaking the nutmeg will indicate whether it is fully ripe: if the seed can be heard tinkling inside, it is time to let it out – and it is this seed that contains the precious spicy aroma, a sweet invitation to travel.

The stripped seeds are dried away from sunlight and turned over every day with a large wooden spatula to avoid any degradation of the lipids, whose fusion temperature is 38 °C. After six to eight weeks, the nut inside the seed (the nutmeg, strictly speaking) is dry enough to be separated from its rock-hard husk, before being crushed and subjected to steam distillation. Six to twelve hours later, a warm, sweet and spicy essential oil is ready. Mace – sometimes incorrectly referred to as the 'nutmeg flower' – is also scented and can be distilled, although its essence is distinctly more bitter than that of nutmeg itself. The two spices are occasionally distilled together.

The facetted smell of nutmeg can be disorientating. By turns hot, cold, fresh, aromatic, woody, burning, peppery, exotic, it has a complexity that can overwhelm our senses. When used in cooking, a pinch of nutmeg adds a warm, flavoursome touch to vegetables au gratin, a pumpkin soup, a bechamel sauce or old-style mashed potatoes, as well as to a compote, custard cream or rice pudding. Similarly, a perfumer does not need to use very much nutmeg essence to benefit from its effect. On the contrary: nutmeg is so dense that it makes its presence felt even in tiny amounts. Yet it shines most when it is combined with other spices rather than being left to its own devices, and invariably ends up establishing a vibrant conversation between the nose and the taste buds.

Genus and species: Myristica fragrans
Family: Myristicaceae
Type: Tropical tree with a pyramidal canopy (10–12 m)
Main growing areas: Indonesia, Sri Lanka, Grenada
Harvest: April, August and December
Part used: Nut
Treatment: Steam distillation (essential oil)
Olfactive family: Spicy
Facets: Aromatic, fresh, peppery, warm, dry, dusty

Osmanthus
China

'The extraction of this deliciously apricoty flower takes place in China, and nowhere else. It is a piece of civilization ushered into a perfume.'

Os-man-thus: this is surely one of the most romantic words in perfumery, evoking journeys to distant lands and the spells cast by their fragrances. These three syllables enjoy an equal emphasis, gliding into each other and sliding down the throat like sweet syrup. They denote a small, unassuming white flower, whose only visual feature of any note is its four-lobed corolla. Its scent continues the romantic theme, however, imbuing the air with a soothing aroma reminiscent of apricot and conveying a sense of foreign climes, bathed in sunshine, peaceful, harmonious and cordial, far from all danger. The osmanthus is a prickly leafed shrub, closely related to the olive tree, and it has never left its native China. To find it, a traveller must first go down the Li River, overhung by extraordinary rock formations dating back many millennia that rear up like a stone forest – a mysterious landscape that is depicted on China's twenty-yuan banknotes. On arriving in Guilin, in the autonomous region of Guangxi, bordering Vietnam, the visitor is immediately intoxicated by the aroma of osmanthus. The whole region is covered with plantations that stretch for hundreds of kilometres, to such an extent that they have given the city its name – Guilin literally means 'osmanthus forest'.

The tiny white flowers blossom in large clusters. The harvest takes place every year over three weeks straddling September and October, although in some years blessed by particularly clement weather there can be an extra florescence in May or June. Groups of women use long bamboo poles to shake the branches vigorously and dislodge the flowers, which are gathered in large white sheets spread under the foliage. This process is not repeated every day and the extraction process does not begin until the end of the harvest, so in the meantime the collected fresh flowers are immersed in brine to preserve them. This salty preparation gives the flowers a leathery, animal smell of varying degrees of intensity, which has been compared to that of olive paste. A perfumer will either embrace this smell or reject it, depending on their taste and the aesthetic concept that they are pursuing. It is a sort of minor miracle to see such a subtle and poetic aroma arise from a technology so basic and old-fashioned: the equipment used for the extraction process looks like an outsize washing machine and dates back to the 1940s.

In China, the fragrant osmanthus flower does not enjoy the same aura of exquisiteness as it does in the West; it is mainly used in infusions and as the primary ingredient in a perfumed tea (*gui hua cha*). The main attractions of the osmanthus for a perfumer are its honeyed floral sweetness and its splash of apricot, sometimes veering towards ripe peach. It is a scent devoid of provocation or any overwhelming sensuality. Its welcoming sweet and fruity note, reminiscent of an easy, familiar laugh, adds a natural touch to any floral bouquet. The velvety softness of osmanthus blossom may bring to mind rose petals, but the delightful paradox of this flower, so unequivocal in its florality, is that it is never treated as a flower by perfumers: as far as they are concerned, it embodies apricot.

Genus and species: Osmanthus fragrans
Family: Oleaceae
Type: Evergreen shrub (3–12 m)
Main growing area: Southern China
Florescence: September–October (less commonly, May–June)
Part used: Flower
Treatment: Extraction with solvents (absolute)
Olfactive family: Fruity
Facets: Floral, honeyed, jammy, fruity (peach, apricot), leathery

Oud Assam

Bangladesh

'The smell of oud essence transports us far back into our animal memory to rediscover who we once were.'

Beauty sometimes emerges in very curious ways in the natural world. Silk thread, for instance, owes its existence to the insatiable appetite of the *Bombyx mori* moth for mulberry leaves. Similarly, the odiferous resin known as 'oud wood' (also called 'agarwood') is secreted in the heartwood of a tree defending itself against an infection.

Our story begins in the jungle of Assam, Bangladesh, the historic cradle of agarwood – uninhabited, apart from a few Bengal tigers. In this remote spot, a parasitic insect attacks the pale bark and smooth trunk of a tree from the *Aquilaria* genus, in order to feed on its sap; this causes an infection in the form of the fungus *Phialophora parasitica*. In response, the tree produces a dark resin that hardens its wood and protects it from the aggression. In this botanical lottery, each individual tree may conceal a black treasure at its heart – or not, as the case may be. This priceless bounty of nature is thus the fruit of the mysterious workings of chance.

Merely detecting infected trees requires considerable expertise, bearing in mind that it takes two to three years for the blackish resin to rise to the surface. Techniques have been developed to control this natural process more efficiently – notably, the insertion of nails in the trunk to provoke a fungal infection. However, the resulting essence is of much poorer quality.

Harvesting involves removing the bark, splitting the wood and reducing it to dry chips over the course of several days, before soaking these in spring water for almost two weeks to make the fragrant resin appear. This resin is then processed: a complete distillation can take up to ten days, resulting in an essence of exceptional quality, with varying degrees of animal or faecal notes, depending on the tree. This essence is extremely expensive (around £35,000 per kilo), because a tree cannot secrete the precious resin at all until it has reached the age of ten years – not to mention the fact that it takes as long to train a good distiller. The years spent waiting are reflected in the price. Furthermore, every oud is unique, as every tree develops its own particular disease, and thus its own noticeably distinct resin. Grades running from one to seven, according to a resin's origins, have been established to differentiate nuances of quality.

One of the main characteristics of this woody note – *al-oud* means 'wood' in Arabic – is its capacity to transport – not just from one place to another but also in time. Oud wood has been used for centuries for medicinal and spiritual purposes. Herodotus recorded its application by Ancient Egyptians in embalming rituals. According to the Song of Songs, it was planted right in the middle of the Garden of Eden. Buddha supposedly called its scent the smell of nirvana, while for the Prophet Muhammad it was the smell of paradise. In effect, it is revered by almost all traditions. The 'wood of kings' also plays a central role in the Japanese *kodo* ceremony, where it is burned and 'listened to' by the participants.

This nugget of perfumery is a gift from heaven that prompted Europeans to rediscover transgressive animal notes, which had been rejected by our prudish noses. And it is true that, on initial contact with galbanum, the nostrils are at first roughly assaulted, before discerning a spicy saffron facet that provides an unexpected incandescent glow; from here it is easy to lose oneself in its mysteries, bewitched by the aura of this divine essence.

Genus and species: Aquilaria crassna
Family: Thymelaeaceae
Type: Tropical tree, or sometimes bush (6–20 m)
Main growing areas: Bangladesh (Assam region), Laos, Thailand
Harvest: All year round
Part used: Wood chips
Treatment: Hydrodistillation or steam distillation of the wood chips (essence)
Olfactive family: Woody
Facets: Woody, smoky, leather, animal, fruity, spicy (saffron)

Patchouli
Indonesia

'Even though it conjures up the mystery of Asia, patchouli has participated like no other raw material in the creation of a typically French-style perfumery.'

Patchouli summons up the humus, forests and damp soil of Indonesia, where it originated. A paradoxical and often misunderstood ingredient, patchouli is the object of many popular misconceptions. It is seen as extremely feminine, but in fact the sensuality that it transmits cannot be confined to any particular gender: it works as effectively in both feminine sandalwood and ambery perfumes as it does in masculine woody and leathery accords. It is considered serious and sombre but – while it is true that it brings a certain depth to a composition – it also has the capacity, even when used only as a trace element, to express the coolness of an eau de Cologne or an eau fraîche.

The patchouli shrub barely reaches a height of one metre and it grows only in a hot and humid tropical climate. Its broad downy leaves – recalling mint leaves – emerge from a long, hairy stem. Its essential oil lurks in all parts of the plant, including the root, but connoisseurs have discovered that the best-quality oil comes from its soft upper leaves and small branches. This oil is obtained by using steam to distil part of the stem and the dried leaves (whose scent is barely perceptible when fresh). The essential oil must be left to rest in a barrel for several months in order to eliminate all traces of bitterness. It is then subjected to a final rectification to dispel any unpleasant mouldy, saltpetre-like facets. The end result is a gentle, tender and luminous heart of patchouli.

This plant's sing-song name is sufficient to evoke foreign climes on its own. *Patchouli* is a compound word in Tamil, bringing together *patch* ('green') and *ilai* ('leaf'). In the nineteenth century, patchouli found its way to Britain, where its aroma contributed to the potpourris used in middle-class homes and permeated the fine linen of the aristocracy. In the 1850s, it began to acquire a more unseemly reputation in France, under the Second Empire. High-class prostitutes developed a liking for splendid cashmere shawls from India that had been wrapped in patchouli leaves (they are highly effective moth repellents). The appeal of these shawls owed as much to their smell as to the cloth itself. The standing of patchouli was dealt a further blow by these women's habit of spraying it generously over themselves. In the 1960s, this wonderful essence suffered further indignities when it was rediscovered by hippies in India, who brought home cheap (and sometimes dubious) patchouli bases. All these developments tarnished patchouli's aura of mystery and elegance. Nowadays, however, this once despised scent has become, along with that of citrus fruits, the principal ingredient of modern perfumery, with its almost perfect blend of rusticity and sophistication. Contemporary perfumers have no intention of forgoing its powerful, organic, camphorous, woody and earthy aromas, complemented by a delightful cocoa facet. The special place that it has earned in the hearts of perfume lovers is beyond all question. Patchouli is probably one of the most potent essences that can be extracted from a plant; it constitutes a whole perfume in its own right, making it a perfect link between humans and the natural world.

Genus and species: Pogostemon cablin
Family: Lamiaceae
Type: Shrub with green leaves (1 m)
Main growing areas: Indonesia (Java, Sumatra and Sulawesi), Malaysia, India, China, Brazil
Harvest: May–September (six months after being planted)
Parts used: Leaf and stem
Treatment: Steam distillation (essential oil)
Olfactive family: Woody
Facets: Earthy, smoky, liquorous, camphorous, musty (cellar)

Sandalwood Australia

'There is a great deal of majesty in this tranquil but dense and generous wood, like a slow, warm piece of music, and it makes a perfume exceptionally soothing.'

The mere mention of sandalwood conjures up the sticks of incense infused with sandalwood paste that burn all day in Hindu and Buddhist temples. It could well be one of the first perfumes ever used by human beings. In India, *Santalum album* originated in the province of Mysore and it has been an integral part of the country's culture since time immemorial. It was, however, overexploited to such an extent that it has almost died out there, and so perfumers had to find another source of comparable quality. This quest eventually led to Australia, which saved this cultural legacy from extinction by becoming the world's leading provider of sandalwood. Production is concentrated in the area around the town of Kununurra, in the north of the country, with a tropical climate resembling that of southern India. This similarity has enabled 'white' sandalwood to adapt to its new setting so successfully that it has now achieved the same level of sophistication and complexity as its Indian counterpart, the famed treasure of Mysore. Australia also boasts another, autochthonous variety, *Santalum spicatum*, which grows in the arid expanses of Southwest Australia; this is markedly more rustic than the *S. album* variety, with a greener head note and earthier, smokier and more resinous inflections.

Before contributing its sacred wood to cremation ceremonies, sandalwood is an aromatic tree that grows to a height of up to 10 metres. It has the distinction of being a hemiparasite: the tree's roots attach themselves to those of a host (preferably an acacia or a eucalyptus), in order to take advantage of its sap and feed on its reserves of micronutrients. This forced marriage also enables sandalwood to protect itself from the sun. As it grows, it gradually accumulates oil, particularly in its roots, but only those specimens that reach maturity (thirty to fifty years, the peak of their aromatic capacity) produce the characteristic smell so coveted by perfumers. A tree that is ready to deliver this olfactory wonder can be identified by measuring the circumference of its trunk: 50 centimetres is the crucial marker. Its exquisite scent lies in the heart of the wood, which is pale green or white (hence its name *album*, meaning 'white' in Latin).

During the rainy season, sandalwood roots are full of essential oil, so this is when the tree is uprooted. Its bark and sapwood – the youngest and softest part – are removed and the remainder is left to dry out. The clear, slightly viscous essential oil, which faithfully reflects its raw materials, is obtained via the hydrodistillation of wood chips and pulverized rootlets. This operation can take between two and five days. The resulting essence is often considered 'feminine', on account of its delicate, creamy and milky scent, but perfumers are reluctant to subscribe to such a gendered classification of ingredients. Sandalwood, that quintessential Indian fragrance, is indeed gentle but it is also equally well endowed in masculine fragrances that present a subtle opulence. The devotion that has been offered to sandalwood for thousands of years shows absolutely no sign of wavering.

Genus and species: Santalum album
Family: Santalaceae
Type: Tropical tree (10 m)
Main growing areas: Australia, Sri Lanka, India
Harvest: All year round
Part used: Wood chips
Treatment: Extraction via hydrodistillation (essence)
Olfactive family: Woody
Facets: Woody, milky, creamy, warm, earthy, (very slightly) spicy

Tuberose
India

'This sensual flower with a smell so dazzling that it is a bouquet in its own right is the absolute embodiment of femininity in all its complexity and humanity.'

The tuberose has acquired a reputation as a seductress. Legend has it that young girls in Renaissance Italy were forbidden to walk in a tuberose garden at night in the company of a young man, in case they fell prey to the plant's narcotic fragrance and, by extension, impulses that contravened their Catholic upbringing. This prohibition reflects another peculiarity of the aroma of the tuberose flower, which is popularly known in India as 'night perfume': it increases in intensity once the sun goes down. Native to the foothills of Mexico, the tuberose was cultivated by the Aztecs 600 years ago, but since the nineteenth century it has particularly thrived (often alongside Arabian jasmine) in the humid heat of southeastern India, especially in Tamil Nadu and Mysore, on the banks of the River Kaveri.

Like a graceful young muse, the tuberose with its long, slender stem topped with trumpet-like white flowers could have been sculpted by Alberto Giacometti or one of his followers. A heliotropic plant (turning in the direction of the sun), it is fragile and requires special attention. In winter, tuberose bulbs must be removed from the ground to protect them from cold and inclement weather. The bulbs are carefully graded according to their size, as only the larger ones will be reserved for flowering – a process that requires a long, patient wait: bulbs planted in spring may blossom the following year, but they can sometimes take another year. Whichever the year, the appearance of flowers in May sparks great activity. The harvest ritual starts at around 4 a.m., when the sun has barely started to rise and the buds are still closed. Picking the flower demands a skilful technique: the floral tube is first gripped between the tips of the thumb and index finger and then peeled downwards. Most of the harvest is immediately transported to markets (the tuberose is primarily an ornamental flower, used to adorn temples and women's hair); the remainder is taken to be processed for the perfume industry (the concrete obtained by this first extraction will undergo further treatment in Grasse).

Tuberose is also traditionally used throughout India to decorate the wedding chamber of newlyweds, as its fragrance is considered aphrodisiac. Yet again, this natural perfume cannot shake off its erotic connotations. The passion that tuberose arouses is not lost on perfumers seeking dramatic effects, although some may lack the mastery or experience needed to keep these effects under control. Descriptions of the far-from-innocent beauty of the tuberose tend to stress its sultry, sensual and intoxicating qualities, setting it apart from other white flowers with which it shares many characteristics, such as jasmine, orange blossom or even ylang-ylang. Tuberose is classified on the perfumery colour chart as 'red' – an incandescent and passionate red, like the velvet of a boudoir.

The most common extraction of tuberose – the absolute – is not only captivating, it is also extremely facetted. Just like the gleaming white flower, it triggers raptures with its successive milky, sunny waves and its green, camphorous and medicinal characteristics – and, of course, its expression of animal sensuality. All these facets combine to make the tuberose a dark, mysterious enchantress.

Genus and species: Polianthes tuberosa
Family: Asparagaceae
Type: Herbaceous perennial bulb plant (1–1.5 m)
Main growing areas: India, France (Grasse), Mexico, Guatemala
Harvest: March–December
Part used: Flower
Treatment: Extraction with volatile solvents (absolute)
Olfactive family: Flowery
Facets: Medicinal, camphorous, green, spicy (clove), creamy, waxy, coconut

Europe

236–241 Calabrian Bergamot, *Citrus bergamia*, Italy

242–247 Blackcurrant bud, *Ribes nigrum*, France

248–253 Citron, *Citrus medica*, Italy

254–259 Rock rose, *Cistus ladanifer*, Spain

260–265 Lemon, *Citrus x limon*, Italy

266–271 Iris, *Iris pallida*, Italy

272–277 Jasmine, *Jasminum grandiflorum*, France

278–283 Lavender, *Lavandula angustifolia*, France

284–289 Sicilian mandarin, *Citrus reticulata*, Italy

290–295 Mimosa, *Acacia dealbata*, France

296–301 Narcissus, *Narcissus poeticus*, France

302–307 Centifolia rose, *Rosa x centifolia*, France

308–313 Damask rose, *Rosa x damascena*, Bulgaria

314–319 Clary sage, *Salvia sclarea*, France

Calabrian bergamot
Italy

'There is nothing unpleasant to the nose in bergamot, nothing that could be considered negative. There are not many raw materials about which that could be said.'

To track down bergamot, head to southern Italy and the scrubland that runs along the Ionian coastline of the province of Reggio Calabria, facing Sicily. Any map of the distribution of bergamot would give pride of place to this strip of land, some 120 kilometres long. It enjoys a microclimate that is hot and humid in summer and rainy in winter, making it an ideal home for this citrus fruit ever since it first took root in this region in 1750.

Despite its unmatched reputation, the origins of bergamot are still a botanical mystery. Its yellow colour suggests that it could be the result of a genetic mutation of the lemon, the bitter orange or the key lime, but this has never been definitively verified. Even the etymology of the word 'bergamot' has itself been subject to various interpretations. According to one theory, the fruit was discovered in the Canary Islands by Christopher Columbus and its name alludes to the town of Berga, close to Barcelona; another line of thinking traces its name back to the Turkish term *beg-ar mudi* ('lord's pear'), in allusion to the fruit's shape. In this case, it would have come from West Asia and been brought back to Europe by the Crusaders.

The bergamot is cultivated principally in order to extract essential oil from its zest. The fruit itself is hardly a delicacy – in fact, its extremely bitter and acidic flesh is fairly unpalatable; it is used only to flavour teas (including the famous Earl Grey) and candies (as in the *bergamotes* typical of Nancy) or, in crystallized form, as an addition to a Moroccan tajine. Bergamots are also used in Calabrian cooking as a wonderful counterpoint to the spicy taste of *'nduja*, a spreadable sausage.

The fruit is gathered from orchards entirely by hand, between January and March, when its skin has yellowed from the effect of the cold. The pickers wear gloves to avoid damaging the precious zest of the four varieties of bergamot, each endowed with an evocative name: *castagnaro*, *feminello*, *calabrese* and *fantastico*.

The trick is to draw from this unappealing citrus fruit an essence so delicate that it provides a poetic substratum capable of brightening up a perfume. This transformation occurs in a fairly rudimentary machine known as a *pelatrice*, which has a series of rollers equipped with sharp points that puncture and scrape the skin. The bergamot is sprayed with small jets of water throughout this process, which wash the skin and conduct the essential oil to a centrifuge, where it is separated from the water. The resulting essence can be green or fruity and floral, depending on whether the bergamot is harvested at the beginning or end of the harvest season. A 'rectified' essence of bergamot, stripped of phototoxic components such as bergapten (which leaves marks on skin exposed to the sun), has gradually replaced the traditional version.

This nectar, unrivalled in its richness and complexity – with 350 chemical components, far more than the essence of any other citrus fruit – has become a universal marker of freshness that continues to fascinate perfumers. An additional reason for its sustained popularity has nothing to do with its olfactive qualities: bergamot has the ability to amalgamate all the raw materials of a formula and therefore serves as a natural solvent.

Genus and species: Citrus bergamia
Family: Rutaceae
Type: Bushy shrub (4 m)
Main growing areas: Italy (Calabria), Ivory Coast
Harvest: January–March
Part used: Zest
Treatment: Cold pressing (essential oil)
Olfactive family: Hesperidia
Facets: Fresh, bitter, zesty, green, floral

Blackcurrant bud
France

'Considered one of the only authentic fruity notes, blackcurrant buds bring a strong natural element to a composition and make an equally strong impact.'

The dry, biting cold of January is not greeted with dread everywhere. This weather is ideally suited, for example, to the emergence of blackcurrant buds in Burgundy, the only region in France to produce them. After a patient wait for the arrival of frost, a blackcurrant plant stirs from the dormant state that it has maintained since losing its leaves. Slightly tapered red buds start to appear, bursting with life, as hard and fat as coffee beans. The perfume industry has focused its attention on the 'Burgundy Black', an extremely old variety with the best aromatic qualities, although blackcurrant plants were originally grown solely for their highly aromatic little berries. They even came to take the place of grapevines in some parts of Burgundy when phylloxera ravaged the area in the late nineteenth century.

Blackcurrants have been grown for their buds for a century or so, and during most of that time they were picked by hand, with the help of an ordinary knife. Mechanization made an appearance only around fifteen years ago, in a drive to combine preservation of the buds' integrity with increased productivity. A specially adapted harvester cuts off all the blackcurrant branches at ground level. A metal mesh is then used to separate the branches from the buds. The branches are put back into the soil, while the buds are kept in cold storage (or frozen) before being sent to Grasse, where they are soaked in an organic solvent to extract their aroma.

Blackcurrants are particularly known for cassis, the liqueur invented in 1945 by Canon Felix Kir, the mayor of Dijon, who came up with the idea of combining crème de cassis (blackcurrant liqueur) with burgundy wine made from the aligoté grape. It was, perhaps, this fusion with alcohol that would cause blackcurrants to find their way into perfumery – although the industry began to show an interest only relatively recently. It was not until perfumers in Grasse considered making use of blackcurrant buds (in addition to the smooth-skinned, dark purple fruits) and, above all, created plantations solely devoted to them, that this product finally took its place in the composition of perfumes. In the 1970s, it began to have a decisive and positive impact, thanks to its crisp, resinous, green and fruity extract with unpleasant hints of transpiration and cat's urine – a strong olfactive marker that enables a perfumer to distinguish it from other smells.

Blackcurrant buds rarely play a major role but instead serve to 'texture' an accord or reinforce the overall impression of a composition's naturalness, similar to blackcurrants' long-standing use as a flavour enhancer in foods. Its forthright and divisive smell (sometimes compared to that of box tree) has its devotees, but also its intransigent detractors. Everybody can agree on one point, however: the absolute of blackcurrant buds is one of the few naturally fruity raw materials in a perfumer's repertoire, and virtually the only one inextricably linked to a single landscape – which makes it even more precious.

Genus and species: Ribes nigrum
Family: Grossulariaceae
Type: Small budding shrub (1.5 m)
Main growing areas: France (Burgundy), Poland, Moldavia
Harvest: December–February
Part used: Leafy bud
Treatment: Extraction with volatile solvents (absolute)
and supercritical CO_2 extraction
Olfactive family: Fruity
Facets: Green, fruity, sulphurous, woody, minty, syrupy

Citron
Italy

'This citrus fruit, with its thoroughly Mediterranean generosity and a sweetness not found in the lemon, spontaneously conjures up joy and sunshine.'

The citron has finally achieved its place in the sun. This large citrus fruit with its knobbly rind (hence its nickname 'Quasimodo'), strong fragrance and very little juice is automatically associated with the lemon – and is, in fact, its ancestor.

Its Italian name (*cedro*) is explained by its skin, which smells remarkably like cedar wood. The citron is not a showy fruit and rarely displays the full extent of its zest. Simplicity is part of its nature, and it cultivates a form of modesty far removed from the bergamot, its companion in the classic formula for eau de Cologne. Yet it has a great deal going for it. Although it is acidic, as part of the great *Citrus* genus, it is far less so than the lemon: it has a sweet touch that makes it cosy and accessible, evoking childhood memories shot through with confectionery. Candied citron is used in some pastries, particularly panettone (a delicious Italian brioche studded with candied fruits and citric zests) and Sicilian cannoli (small, deep-fried wafers).

Botanists may call the citron *Citrus medica*, but the inhabitants of Cap Corse in northern Corsica – who grow it solely to make *cedratine* (a citrus liqueur) and candied fruit – have christened it the 'golden apple' on account of its bright golden zest – and, perhaps, the wealth that it has brought to the region.

The citron's extreme sensitivity to the cold – unusual for a tree that originated in the southern valleys of the Himalayas – has confined its cultivation to areas with mild winters, such as Corsica and Greece, but above all Calabria and Sicily, especially the Catania region, whose soil has been made fertile by Mount Etna. Here, the citron really comes into its own.

The citron tree looks somewhat flimsy and is modest in size (four to five metres in height); it does not exist in the wild and is not the most charismatic of perfume trees – in fact, it displays an enigmatic modesty. All year long, it produces extremely fragrant flowers – purple or white, depending on whether the fruit is sweet or acid – that are larger than those of any other citrus tree. The rough-skinned citron fruits are equally imposing, weighing up to two kilos and reaching 30 centimetres in length. Immediately after being harvested, they are taken for processing in a *pelatrice*.

The resulting pale-yellow essence resembles that of lemon, although it is sweeter, more delicate and 'warm'. Citron imbues a perfume with a striking freshness and tonicity, as well as – more surprisingly – a glorious florality. This ancestral citrus fruit is endowed with all the elegance of the bergamot, without any of its bitterness. Any perfumer in search of a natural head note, or an arresting splash of yellow, will instinctively be drawn to the citron.

Genus and species: Citrus medica
Family: Rutaceae
Type: Evergreen shrub (3 m)
Main growing areas: Italy (Sicily and Calabria), France (Corsica)
Harvest: Late October–early January
Part used: Zest
Treatment: Extraction by cold pressing (essential oil)
Olfactive family: Hesperidia
Facets: Floral, fruity, acid, green

Rock rose
Spain

'This wonderful amber plant has passed through the ages, and it provided perfumers with their first leather note.'

This plant's penetrating smell evokes the sun-scorched shrublands of the garrigue and the Corsican maquis. The rock rose (sometimes called the cistus) thrives on sunshine and blossoms most fully in the suffocating temperatures of lands devoid of shade. So it is highly suited to the arid soils of the Mediterranean region, with southern Spain being its clear favourite. In western Andalusia, between Seville and Portugal, the hills of El Andévalo, famous for their centuries-old cork-oak forests, are covered with thousands of hectares of wild rock rose, a small shrub that can colonize even the most hostile environments – it is a pyrophile, meaning it has the capacity to regrow after a fire – to proclaim its unique fragrant message.

Pretty, delicately wrinkled, white or pink rock rose flowers appear in April but survive for barely two days. Each flower has five petals with a deep red spot, hence its popular Spanish name of 'Christ's tears' – the bloom is embroidered on the Virgin's dress for celebrations held in Andalusian villages in July. Although the flower is undeniably visually stunning, it does nothing at all for the nose, as it does not give off the slightest smell. The fragrance has a different source: when it starts to blossom, the plant secretes a sticky resin that protects it from the sun and prevents dehydration. And the more the temperature soars, the more its stems turn red thanks to the formation of this gum. Its amber, caramel smell – as enveloping as it is intoxicating – pervades the whole area.

In former times, Cretan shepherds used to gather the strong-smelling resin by combing it out of the fleece of their goats, where it would stick after they had been grazing in rock rose fields. The shrub's branches would also be 'whipped' with a *ladanisterion*, a kind of rake equipped with long thongs that served to release the precious resin. Nowadays, the stems are simply cut off with a sickle, confining the harvest to that year's shoots – the tender tips of the branches, where most gum is concentrated. This is exhausting work when the temperature rises to 45 °C. Rock rose gatherers can be readily identified by the unmistakable smell of the resin, which impregnates their clothes throughout the entire harvest season.

The complex aroma of this gum manifests itself in several ways, as reflected in the various terms used in suppliers' catalogues, which can often lead to confusion. Products derived from the distillation of the branches bear the designation 'cistus', while those derived from gum are designated 'labdanum'. Branches are steam distilled to produce a zesty, orangey essence resembling incense, while extraction results in a leathery, amber cistus absolute. When gum is the starting point, the labdanum absolute extracted by volatile solvents is more amber-vanilla. Consequently, this plant is particularly cherished by perfumers because it constitutes one of the few raw materials from the plant kingdom capable of expressing all these ambery and animal smells, along with aromas of grass, wood and medicinal resin.

Genus and species: Cistus ladanifer
Family: Cistaceae
Type: Evergreen shrub (0.3–1 m)
Main growing areas: Spain (Andalusia), France (Corsica), Portugal, Morocco
Harvest: June–October
Parts used: Leaf and stem
Treatment: Steam distillation (essence) and extraction with solvents (absolute)
Olfactive family: Amber
Facets: Resinous, amber, woody, medicinal

Lemon
Italy

'The stunning vivacity of this simple citrus fruit, cool and acidic, immediately transports you to a place under the Mediterranean sun.'

This fruit evokes the air of a sunlit Eden. It was once thought that the famous golden apples in the garden of the Hesperides in Greek mythology could have been oranges and lemons – the hesperidia family embraces the notes of citrus zests in perfumery, including the lemon. There is no truth in this, but what is certain is that lemon trees served as decorative trees in the traditional Islamic garden, which was an allegory of an ideal, heavenly world that appealed to all the senses. The lemon automatically conjures up the Mediterranean, with its blue skies, sunshine and *dolce vita*, but, in fact, its origins lie in Asia: it first appeared some 3,000 years ago in the Kashmir region, between India and China. In the tenth century, the Arabs brought what they called the *li mûm* to the entire Mediterranean Basin, starting in Spain, where it became *limón*, which subsequently gave rise to the English word 'lemon'.

The lemon tree grows fast, presenting dense, irregular foliage, and it blossoms several times a year. It also generously supplies fruit all year round, although the various waves of fructification can vary in intensity. The lemon tree flourishes in sunshine in any setting, and is vulnerable to cold. It is ideally suited to the dry, warm climate of Sicily, especially the regions of Palermo, Messina, Syracuse and Acireale (where the lemon-based liqueur limoncello was created). It thrives on this area's fertile soil close to the seashore, but is never found above an altitude of 300 metres.

Lemons are ripe when the peel turns completely yellow. They are tightly packed on the tree and can be gathered only by hand, by cutting the stem with specially designed palm-sized shears. After starting on the ground, lemon pickers end up standing on a ladder, sometimes contorting themselves to reach the top branches of a tree that can grow to a height of six metres. The fruits are then washed in a tank before the essence is extracted via mechanical cold pressing of the zest – rather than, as might be expected, via steam distillation – in a machine known as a *sfumatrice*. (The term *sfumatrice* is thus used in the designation of origin of lemon essence, rather than that of its essential oil.) Sponge extraction, which is much more labour-intensive, has virtually disappeared. This rudimentary technique involved pressing the skin of a half-lemon (with its pulp previously removed) against a sponge to painstakingly collect its elixir. Lemon essence presents volatile notes that are fresh, zesty, clean, acidic and lively, making it an ideal basis for an eau de Cologne or an eau fraîche, or simply for injecting a touch of healthy naturalness and light into a composition.

Genus and species: Citrus x limon
Family: Rutaceae
Type: Evergreen tree (4–6 m)
Main growing areas: Italy (Sicily), Spain, Argentina, Brazil
Harvest: November–May
Part used: Zest
Treatment: Cold pressing (essence)
Olfactive family: Hesperidia
Facets: Fresh, acidulated, acidic, zesty, exuberant, juicy

Iris
Italy

'The iris needs time to reveal its treasures – that is part of the nobility of this olfactory miracle, able to set the heart of a perfume racing.'

This flower is set apart right from the beginning by its haughty demeanour and the fact that it is named after a goddess – in Greek mythology, Iris was sent by the gods to deliver messages of hope to human beings. Its sophisticated beauty hides a secret, however, in the underworld of its roots, as its smell lies in the heart of its rhizome, a subterranean stem that sprouts buds. This rhizome has to be patiently allowed to offer up its aroma in its own time but, ultimately, it is worth its weight in gold.

After a wait of three years, the rhizome is dug up, cleaned and then stripped of its rootlets, before being peeled like a mushroom with a small, curved knife called a *roncolino*. No machine has proved more effective than the human hand in harvesting this buried treasure, as it is often burrowed inside awkward crevices. The harvest is merely the start of a process that will take years. At this point, the unsightly rhizomes barely smell at all. First left out to dry in the sun on raised racks, then packed into jute bags, they will be stored for a further two to three years, to allow the fragrance to develop. Over the course of this 'desiccation' period, magic comes into play, as, quite naturally, one by one, the precursors of irone (the active principle that gives the iris its smell) begin to oxidize, unleashing all their rich aromas, particularly that beautiful scent of hazelnut and precious wood, a veritable olfactory masterpiece. After these years of storage, the rhizomes, by now as hard as rock, are reduced to powder in a grinder before they can be processed.

The Italian variety of the plant (*Iris pallida*) has by far the highest irone content and is thus the most highly prized by perfumers. It grows in an area between Florence, Siena and Arezzo, shaded by olive trees, in the poor, rocky soil typical of the chalk hills of Tuscany. Every May and June, the iris fields in the valleys are awash with the exquisite pale blue that gives the plant its name (*pallida* means 'pale'). The flower, which serves no purpose in perfumery, is picked straight away, to allow the plant to devote all its energy to strengthening its precious rhizomes rather than producing seeds.

The iris has been used for its scent ever since the Renaissance – Catherine de' Medici, for example, perfumed her gloves with it. In those days, the rhizomes were pounded and sieved, before being mixed with rice powder, resulting in a sweet aroma reminiscent of violet. It is for this reason that the notes of iris are classified as 'powdery', but this is an overly reductive description. In reality, its fragrance presents a rare complexity: dry, woody, and cocoa-like (via the resinoid, rather than the essence), with a slight 'raspberry' fruitiness discernible if the nose lingers long enough. The extract also recalls carrot, violet and mimosa. Above all, however, the iris aroma evokes the soil of Tuscany, from which it draws its strength, biding its time amid the general indifference of a world in a hurry. *Iris pallida* serves as a symbol of top-end perfumery stubbornly intent on taking its time.

Genus and species: Iris pallida
Family: Iridaceae
Type: Perennial plant (0.6–1 m)
Main growing areas: Italy (Tuscany: *I. pallida*), France and Morocco (*I. germanica*)
Harvest: July–September
Part used: Rhizome
Treatment: Hydrodistillation (concrete or 'iris butter') and extraction with volatile solvents and alcohol (absolute or resinoid)
Olfactive family: Powdery
Facets: Powdery, woody, straw, violet, cocoa

Jasmine
France

'This flower with its incomparable fragrance – flowery, fresh and expansive – provides the benchmark by which all others are measured, gauged and defined.'

It is said in Grasse that the rose gives its soul to the sun and jasmine gives its soul to the stars. It is a flower that opens only at night, withdrawing its heady perfume when dawn arrives. Therefore, as many petals as possible have to be picked in just three or four hours, before the sun burns the flower and alters its aroma. The picking season lasts for around ninety days, from late July to October, and requires considerable manpower. The jasmine harvest cannot be mechanized, as this small white star-shaped flower is far too fragile to be handled by anything other than delicate human hands. It is a painstaking process, with each flower being picked individually off its stalk to prevent its petals from dropping off. It is also important not to remove any stems or buds, to avoid compromising the following day's blooms. The pickers (usually women) place the jasmine in their aprons, making them look 'pregnant' with flowers by the end of their shift. Even when the weather is hot and humid – ideal for jasmine flowers – these workers can collect up to four kilos of petals in one session.

The *jasmin de pays* (local jasmine) that is so celebrated in Grasse is not, in fact, endemic to the area at all. It originated in the mountains of northern India, from where it was taken to North Africa and then Spain by the Moors, as indicated by its name: *yāsamīn* means 'flower' in Arabic. Some Spanish sailors were entrusted with taking it to Provence – hence the popular name of 'Spanish jasmine'. And ever since it became the glory of Grasse, it has been known there as simply 'the flower'. It owes its ability to survive the low winter temperatures in France to the grafting of the Indian variety on to a common jasmine, which is hardier and more resistant.

For many years, the production of jasmine in gardens was enough to satisfy perfumers. In 1860, jasmine plantations made their first appearance, when the opening of the Siagne Canal guaranteed the irrigation crucial to the plant's growth beyond the city limits of Grasse. By the 1930s, jasmine fields stretched as far as the eye could see around Grasse, creating a floral carpet that extended from Vence (Alpes-Maritimes) to Seillans (Var). Women and children, and on Sundays even a few men, used to go out to pick the flowers, in order to earn a crust or supplement a meagre income, or simply to enjoy the company. Nowadays, only a handful of growers continue this tradition, and on much smaller plots of land, although jasmine flowers are still processed in Grasse.

Jasminum grandiflorum yields no essence, as steam distillation is unable to draw out all the flower's olfactive subtlety. Its perfume can be revealed only through extraction. Over a period of seven hours, three successive baths enable jasmine to macerate and release its fragrant sap into the solvent. A concrete – an aromatic paste – is obtained by evaporating the solvent. Finally, the absolute is attained after dilution in alcohol and filtration. This is an extremely solid extract, less green and less medicinal than the flower itself, with a note of blackcurrant (a sign of quality typical of Grasse). The jasmine absolute derived from CO_2 extraction, a speciality of Louis Vuitton, provides an end result closer to the flower on the plant and represents science's homage to this marvel with its exquisite nocturnal sigh.

Genus and species: Jasminum grandiflorum
Family: Oleaceae
Type: Bushy shrub (1–1.5 m when cultivated)
Main growing areas: France (Grasse), India, Egypt, Morocco
Harvest: July–October
Part used: Flower
Treatment: Extraction with volatile solvents and supercritical CO_2 extraction, developed exclusively by Firmenich for Louis Vuitton (absolute)
Olfactive family: Floral
Facets: Floral, petalled, airy, fruity, animal

Lavender
France

'Fine lavender, with its concentrated aromatic freshness, provides the perfect balance between woody facets and a beautiful florality.'

This plant's name alludes to a custom of Ancient Rome whereby bath water was scented with bunches of lavender – the Latin word *lavare* means 'to wash'. The city's most dedicated followers of fashion wore a perfume called *nardinum*, based on myrrh and lavender. Roman invasions were responsible for the establishment of lavender in Provence, along with that of the vine and the olive tree. Until the nineteenth century, lavender was picked only in the wild. It was not until around 1890 that the first organized lavender plantations began to appear, creating a landscape worthy of a postcard, with great streaks of purple climbing the hills or stretching to the horizon, interrupted only by the immensity of the blue sky. Nowadays, Provençal lavender fields are largely confined to three *départements*: Drôme, Vaucluse and Alpes-de-Haute-Provence. They are always found at an altitude of at least 800 metres, normally on the slope that receives the most sunshine.

Tourists wandering the country lanes of Provence, the sun beating down on their heads, may sometimes think that they are smelling the scent of lavender when in fact the fragrance belongs to its close relative, lavandin, a hybrid of fine lavender and spike lavender. When the two plants are seen together, however, there is no confusing them: lavender flowers are not homogeneous in colour, ranging from blue to mauve with a little grey thrown in, and they have a single, fairly short flower spike. Similarly, scent-wise, true lavender is more sophisticated and delicate, and far less camphorous, than lavandin.

Of all the different varieties of lavender, expert perfumers invariably opt for the precise and noble fine lavender (*Lavandula angustifolia*), with a particular predilection for that of Haute-Provence, which enjoys a Protected Designation of Origin (PDO) that vouches for its outstanding aromatic qualities.

Lavender is harvested at the height of summer because high temperatures encourage the essence to rise up in the plant's secretory glands. The sprigs are particularly fragrant if they are gathered immediately before the flowers open. Some perfumers allow the harvested lavender to dry out for two days prior to the distillation process, in order to eliminate any excess moisture and ensure the integrity of the essence. Essential oil is diffused throughout the plant, apart from in the roots, and is extracted by hydrodistillation of the tips of the flowers.

Lavender brings a note of joyful freshness to even the most arid landscapes by swathing rocky hillsides in purplish blue. It has exactly the same effect in a flacon: its radiant, aromatic note recalls gardens in the South of France in the summer holidays, as well as old-fashioned shaving soap and the delicacy of a whole art of living embodied in the lavender sachets that our grandmothers used to tuck into their linen cupboards. A perfumer will tend to highlight its grassy, floral, medicinal and camphorous notes (although the latter are more pronounced in lavandin), resulting in a vintage feel that brings to mind the smell of eau de Cologne.

Lavender also exhibits qualities that go beyond its powerful scent, as its frequent use in aromatherapy has heightened its capacity to project an image of relaxation and well-being. As the German psychologist and author Heike Ullmann observed so charmingly: 'The colour of calmness is mauve, and its perfume is that of lavender.'

Genus and species: Lavandula angustifolia
Family: Lamiaceae
Type: Perennial evergreen shrub (2 m)
Main growing areas: France (Provence), Bulgaria, Ukraine, Moldava, China
Harvest: Late July–late August
Part used: Flower
Treatment: Steam distillation (essence)
Olfactive family: Aromatic
Facets: Floral, grassy, clean, camphorous, hay

Sicilian mandarin
Italy

'Its essence smells of vitality and glorious sunshine, bringing to mind the scent left on the hands after enjoying a segment of mandarin.'

It cannot be denied that, fairly or unfairly, the mandarin orange does not enjoy the same reputation as the bergamot, which is generally considered richer, classier and more select. Nevertheless, in terms of fragrance, this sweet-tasting festive delicacy is one of the most 'talkative' of all citrus fruits, as well as one of the most nuanced.

The essence of mandarin uncannily resembles the fruit that we all know and love. Jacques Cavallier-Belletrud decided to make it the star of *Jour se lève* because he recognizes its olfactive range better than anybody. He found within the essence of mandarin facets of orange blossom and of orange itself, of lemon, grapefruit and even thyme (thanks to the presence of thymol). The mandarin has great evocative powers: its fragrance sparks emotions and memories, conjuring up a journey through a Mediterranean Garden of Eden or childhood recollections of opening presents under the Christmas tree. It has the added advantage over other citrus fruits of avoiding overly direct and somewhat retro references to eau de Cologne.

The mandarin tree – a small, sturdy shrub with a round canopy and bright green lanceolate leaves – came originally from China, where it has been grown for thousands of years. It even sports the same colour as the robes of mandarins, high-ranking officials of imperial China. This coincidence explains how the fruit received its name, coined by the Portuguese, who took it to Europe in the early nineteenth century. Calabria proved to be its ideal breeding ground, particularly the Tyrrhenian and Ionian coastal regions (Reggio Calabria) home to most of the area's citrus orchards (known as *giargini* in the local dialect). The mandarin trees here are protected from the wind by hedges of olive trees. The fruit starts growing in the summer and ripens from October onwards, depending on the variety. It should be noted that mandarin blossom is completely ignored by perfumers, in favour of the rough-skinned fruit.

The mandarin harvest is traditionally divided into three parts: October, November, and February and beyond (corresponding to green, yellow and red mandarins respectively). The three resulting essences are extremely different in terms of their fragrance. The first, the green essence, is particularly fresh and potent, with zesty, herbaceous and mineral elements; the yellow essence is more delicate and fruity, while its red counterpart is much sweeter and juicier.

Like all citrus fruits, mandarin is extracted via cold pressing, a method that was first introduced into southern Italy and Sicily in the nineteenth century. The older manual technique of the 'sponge' – which involved pressing the fruit, minus its pulp, against a large sponge to draw out its essence – has virtually disappeared. These days, two pressing techniques are generally used. The first involves putting the fruit into a *pelatrice*. This technique coexists with the *sfumatrice-torchio*, whereby the entire mandarin is squeezed and crushed to extract the essence from both the zest and the juice.

The mandarin's floral, slightly orange blossom-tinged facet, and all its aromatic tones (thyme) are miraculous. A perfumer can play with this complexity, as well as with a particularity of this winter fruit with the summery taste – a gourmand sweetness not found in other citrus fruits. Although it is not formally classified in perfumery under fruits (unlike strawberry, peach and banana), the mandarin has all the mouth-watering brightness of other fragrances that are recognized as such.

Genus and species: Citrus reticulata
Family: Rutaceae
Type: Evergreen shrub (3–6 m)
Main growing areas: Italy (Calabria, Sicily), Spain, Morocco, Brazil, Argentina, China
Harvest: October, November and February
Parts used: Whole fruit and zest
Treatment: Cold pressing of zest or whole fruit
Olfactive family: Hesperidia
Facets: Fresh, sweet, acidulated, candied fruit

Mimosa
France

'The absolute of mimosa gives a floral bouquet its green character and a breath of naturalness. It is like a smile in the middle of winter.'

Mimosa's yellow pompoms provide relief from the dour solemnity of winter, which doubtless explains its pre-eminence on the Côte d'Azur – in fact, mimosa has become one of the region's emblems. It originally arrived there from the Antipodes. Seeds of *Acacia dealbata* were first brought back from Australia and Tasmania by the explorer Captain James Cook in 1771 and sown in Kew Gardens, London. The resulting plants withered in that forbidding climate, and it was not until wealthy Britons developed a passion for the Côte d'Azur that, around 1850, mimosas were imported to embellish the gardens of the newcomers' sumptuous villas. Eventually, Joséphine de Beauharnais, Napoleon's first wife and a lover of exotic plants, brought this little yellow flower to the attention of the French. She planted mimosa in her large hothouse in the gardens of Malmaison, outside Paris, while also having it included in a work published by the botanist Étienne Pierre Ventenat. It is partly thanks to her, therefore, that this sun-loving shrub now grows on the slopes of the Tanneron Mountains, in the Var region, surrounded by eucalyptus trees, forming the world's largest expanse of mimosa, spread over a golden triangle embracing the towns of Mandelieu, Tanneron and Pégomas. Indeed, this exotic plant's integration into the landscape has been so complete that it has even given its name to the village of Bormes-les-Mimosas. It quickly adapted to the area's siliceous soils and now thrives in its sunny climate, only rarely disturbed by frost.

This 'winter sun' brightening up the heights of the Estérel Massif became known to walkers for its visual exuberance before it began to tickle the noses of perfumers from Grasse. However, anybody wandering through these hills in the month of February can experience the olfactory effect for themselves – they have only to close their eyes and drink in with every pore of their being that refreshing, powdery smell, borne on the gusts of the mistral. A smile will instantly spring to their lips – an involuntary smile straight from childhood.

The mimosa's tiny flowers have long yellow stamens but no petals, and they cluster together to form fluffy pompoms. All the perfume is concentrated in these flower heads, from which individual flowers are extracted. Mimosa '*pays*' absolute is more highly prized for its vegetal facet (cucumber and violet leaf) than for its powdery note – an authentic powdery note is best served by cassia flowers (*Vachellia farnesiana*). Mimosa sets off other flowers in a bouquet by bringing out their specific qualities; it can, for example, enhance all the facets of rose or intensify the natural qualities of jasmine. Even an inexperienced nose may recognize mimosa's almond and aniseed touches, but a seasoned perfumer can also detect hints of honey and straw, as well as mastic and linseed oil. Mimosa is an 'impressionist' raw material that enables a perfumer to convey a gentle, muted impression of 'floral snow'.

Mi-mo-sa. These three syllables are synonymous with sunshine, warmth and mellowness. Thanks not only to its fragrance but also to its symbolism and cheery optimism, this flower is perfectly suited to our times. Down with all those naysayers who so hastily banish it to the cosy nostalgia of perfumes from the 1950s to the 1970s! Mimosa belongs resolutely to the here and now.

Genus and species: Acacia dealbata
Family: Fabaceae
Type: Evergreen shrub (3–10 m)
Main growing areas: France (south), India, Morocco
Florescence: Late January–March
Part used: Branch in bloom
Treatment: Extraction with solvent (absolute)
Olfactive family: Powdery
Facets: Warm, green, powdery, honeyed

Narcissus
France

'This fragile-looking flower growing on primordial volcanoes reveals its wild, sensual nature in its perfume.'

When spring finally takes hold on France's Aubrac volcanic plateau between Lozère and Cantal, a curious phenomenon occurs: a snow-like carpet suddenly covers the high pastures. Closer inspection reveals a small, fragile flower with pure white petals set off by a central eye fringed with red. Legend has it that the narcissus always looks downward in memory of the beautiful hunter from Greek mythology, Narcissus, who sat by the edge of a pool and gazed into the water to admire himself.

Since time immemorial, *Narcissus poeticus* – 'the poet's narcissus' – has grown in the wild in this region at altitudes above 1,000 metres. It flourishes on the damp mountain pastures, and its habitat has been preserved in the fields where hay is grown, an ideal environment that allows it to blossom in all its glory. There is no other perfume flower capable of flourishing in this way without any human assistance or intervention. Nevertheless, the yields of *Narcissus poeticus* can vary enormously from one year to the next, and neither farmers nor botanists have been able to work out the reason for these discrepancies.

A strong smell of hay carried by the icy wind on the upper slopes indicates that the narcissus harvest can begin, just a few days after that of its relative, the daffodil. The days of picking narcissus by hand are long gone, replaced by a slightly less rudimentary and onerous – and more productive – semi-mechanized technique. Pickers are equipped with a rake some 30 centimetres wide, mounted on two bicycle wheels, enabling them to remove narcissus heads without damaging the stems of the grasses growing around them. This tool was custom-made for the local conditions, and it permits a single picker to collect 20 to 25 kilos of narcissus per day. The harvesting is usually confined to three short weeks from late May to mid-June, before livestock is taken to graze in the upper pastures. The flowers should be gathered just when the buds start to open, to prevent any dissipation of their fragrance. Once a picker's rake is fully loaded, the narcissus flowers are transferred to a hessian sack and weighed, right there in the field, with a steelyard balance. The harvested crop is taken straight to the distillery. The flowers are first aired, to avoid any fermentation, and then placed in a receptacle filled with solvent; the extraction process lasts around four hours.

Auvergne and its volcanoes endow the narcissus with a crisp, green, somewhat austere earthiness. This type of lush, assertive naturalness has always been a boon to committed perfume lovers. The absolute of narcissus, a hyper-concentrated extract of the flower, emphasizes still further its strident, acrid green notes, along with a leathery animal touch that recalls, in some respects, the smell of a stable. Every time a perfumer decides to pay tribute to this wild, rustic beauty drawn from the well-preserved habitat of Aubrac in one of their compositions, they are also throwing the spotlight on an exceptional natural legacy.

Genus and species: Narcissus poeticus
Family: Amaryllidaceae
Type: Perennial herbaceous bulb (30–40 cm)
Main growing areas: France (Auvergne), Morocco, Netherlands
Harvest: May–June
Part used: Flower
Treatment: Extraction with volatile solvents, prior to washing in alcohol (absolute)
Olfactive family: Floral
Facets: Floral, green, full-bodied, spicy, grassy (hay), honeyed, leathery

Centifolia rose
France

'I always feel emotion whenever I smell that warm, powerful fragrance, infused with honey and studded with pepper, but without the slightest animal inflection.'

This is an unusual rose that blooms only fleetingly, a rose rooted in the Grasse terroir, whose fragrant mystery is preserved in the remarkable profusion of its petals. It is this very characteristic that gave rise to its botanical name of *centifolia* ('a hundred leaves'), coined by the Swedish naturalist Carl Linnaeus in 1753. The *centifolia* has a complex, indeed convoluted genealogy. It probably originated in the Caucasus, but the variety that we know today seems to have appeared in the seventeenth century as a result of hybridization performed by a breeder in the Netherlands – which would explain why it features so prominently in contemporary paintings by the Dutch artists Jan van Huysum and Cornelis van Spaendonck. More specifically, the *centifolia* is the end product of a skilful cross between a variety of *Rosa gallica* and *Rosa* x *damascena* (Damask rose), *Rosa canina* (eglantine) and *Rosa moschata* (musk rose).

The *centifolia* played a major role in building the reputation of Grasse (Alpes-Maritimes), having been cultivated in the area since 1650 – although at that time it served to cover up the smells of the local tanneries rather than provide an ingredient for perfumes applied to the skin. The entire annual production is painstakingly picked by hand, flower by flower, only in the month of May (hence its nickname 'May rose'; it is also known as 'cabbage rose', thanks to its multitude of petals). Any flowers left unpicked are totally lost to perfumery, as their petals wilt in the sun and their fragrance quickly fades. The harvesting day begins at 7 a.m. By then, the fields are already awash with the warm, honeyed scent that is the coveted hallmark of the *centifolia*. The pickers (usually women), protected by broad-brimmed straw hats, traditionally divide into pairs, in order to gather the elegant flowers more efficiently: they position themselves on either side of a rose bush, thereby reducing the risk of overlooking even a single flower. With a sharp flick of the wrist they pick the flowers that are barely in bloom, breaking them off with meticulous precision just below the calix but avoiding any pulling motion.

The extraction process takes place in situ, immediately after the harvest, to prevent any fermentation that could alter the delicate perfume. The peduncles, petals and pistils are all tossed into a large extractor and washed several times in a volatile solvent that gradually becomes saturated with the flowers' smell. The resulting concrete, a vegetal jelly that hardens as it cools, is then treated with alcohol to obtain the absolute, which is an exceptional extract with a sweet scent of Turkish delight. No essential oil is obtained from this rose, however, as it yields little or nothing with traditional steam distillation.

The incalculable value of the May rose derives, above all, from the impossibility of reproducing its sweet, smooth, 'petalled' and slightly peppery aromas in a laboratory, however much a chemist might be able to identify each of its components. Furthermore, the decision to add an absolute of Grasse rose to a perfume is an act of daring, as a very precise quantity is required for a connoisseur to discern the desired sign of quality and class. Only a very few perfumers can accurately gauge this amount.

Genus and species: Rosa x centifolia
Family: Rosaceae
Type: Deciduous shrub (1.5 m)
Main growing areas: France (Grasse), Morocco
Harvest: Three weeks in May
Part used: Flower
Treatment: Extraction with volatile solvents and supercritical CO_2 extract (absolute)
Olfactive family: Floral
Facets: Spicy, herbaceous, honeyed, leathery

Damask rose
Bulgaria

'All the facets of this incarnation of femininity abound in soft gentleness, without the slightest aggressive or jarring animal note.'

Its name has already passed into legend. 'Damask' (*Damascena* in Latin) refers to Damascus, the Syrian city where this rose was born. Indeed, having been grown in Damascus since the sixth century, it has become the symbol of the city. The Persian doctor and philosopher Ibn Sina (Avicenna) was responsible for the flower's first distillation, in the eleventh century, but its reputation spread across the wider world only after French knights came to Damascus during the Crusades and fell under its spell. One Robert de Brie was so seduced by its beauty and heady fragrance that he brought a Damask rose tree back to France in 1254 and planted it in the gardens of his château in Champagne.

In the seventeenth century, the *Rosa* x *damascena*, the 'rose with thirty petals', underwent a revival, this time in Bulgaria. The Ottoman Empire was unwilling to depend entirely on Iran for its production of rose water and so began to grow Damask roses around Kazanlak, in the foothills of the Balkan Mountains. Little has changed since then in Rose Valley, which is now more than 80 kilometres long. Although the plant gradually spread to Turkey in the nineteenth century, and then to the Dadès Valley in Morocco, the foremost producer of the Damask rose remains Bulgaria, where its cultivation is very much a family affair. It can present varying degrees of fruitiness, spiciness and lemon notes, depending on where it is grown, while still retaining its essential characteristics.

The Damask rose is traditionally picked early in the morning, when its buds have barely opened and the dew on its petals intensifies its fragrance. The bushes are fairly tall and offer a generous supply of flowers, requiring intensive labour during the three-week harvest period. The picked flowers are put into hessian sacks, which, when full, are taken to collection points prior to the extraction process. Traditionally, the petals were left to ferment before being processed, but this is no longer the case today. Five tons of petals, or three to four million flowers, are needed to obtain a single kilo of essential oil, meaning that they constitute a top-end raw material.

Rosa x *damascena* is the quintessence of femininity and is by far the most widely used rose in perfumery, where its incomparable sophistication never fails to cast a spell. Its 275 components – some are trace elements, but still have a decisive impact on the aroma – defy all attempts to correlate it with any other natural raw materials. Whether one prefers the essential oil with its fruity, or fruit alcohol (marc de Bourgogne), character or the absolute (a deeper, honeyed and more opulent version), the Damask rose offers infinite possibilities to a creative perfumer. One option is to give it a major role and build a bouquet around it; alternatively, it can be placed in the background, allowing it to contribute its richness while remaining indistinguishable from the other ingredients in the composition. Eternal queen and faithful muse of the perfumer, the Damask rose still has plenty of stories to tell in countless tones and languages.

Genus and species: Rosa x damascena
Family: Rosaceae
Type: Bushy shrub (0.5–1 m)
Main growing areas: Bulgaria, Turkey, Morocco
Harvest: May–June
Part used: Flower
Treatment: Steam distillation (essence) and extraction with a solvent (concrete and absolute)
Olfactive family: Floral
Facets: Fresh, fruity (lychee, raspberry), sweet, honeyed, almond

Clary sage
France

'Its note – at once cool, mineral, amber and tobacco – embodies all the nobility of the aromatic notes.'

For centuries, sage was a mainstay of the kitchen garden, prized for its antiseptic and digestive properties. It was considered a panacea – its very name is derived from the Latin word *salvare*, meaning 'to cure' – and was recommended in particular for ulcers and period pains, as well as for alleviating anxiety and soothing insect bites. Little attention was paid to its scent – yet how good it smells! Sage conveys a hint of a sea breeze, a clean, luminous fragrance that is a little woody but also a little spicy. The full effect can be properly appreciated only by taking a trip back to childhood and picking a leaf – a leaf 'as downy as a hare's ear', according to the writer Colette – and crumpling it in your hand.

Sage is sometimes known as *toute-bonne* ('all good') in France, as it grows almost anywhere. It is just as happy in Russia as in the United States, but it is most at home in the French region running from southern Drôme to the Valensole Plateau in the Alpes-de-Haute-Provence. Clary sage has fully adapted to this dry and sunny territory, to such an extent that perfumers value the local sage as a unique and incomparable strain of the plant. Sage is easy to find in these parts: a hiker has only to head towards the long, straight lavender fields on the high mountain slopes, as sage never grows too far away from its aromatic cousin and starts to bloom just a few days before it. Individual sage plants can then be identified by following the carpenter bees that descend on their spikes of bluish-purple flowers to gather nectar.

Of all the host of varieties to be found, only clary sage (*Salvia sclarea*) is used in top-end perfumery, along with, to a lesser degree, common sage (*Salvia officinalis*) – although the latter's more aromatic and resinous elements make it better suited to functional fragrances.

At the start of summer the plant's square stems are cut off by a harvester while still in bloom and collected. The secret to obtaining an exceptional extract is to know that the olfactive profile of sage that is cut, crushed and distilled while still green is considerably less rich than that of sage that is left to wilt in the field for a few hours before being distilled. Perfumers make use of both the essence and the absolute of clary sage. The essence has a cool fragrance with a strong element of bergamot (linalyl acetate) and inflections of tea, while the absolute (which has a 60–70 per cent sclareol content) is heavier, more tobaccoey and herbaceous. Sclareol is the precursor of ambroxide, a synthetic molecule used in perfumery for its amazing ambergris facets, and it has thus converted sage into a bridge between the natural and the synthetic.

Clary sage has finally come into its own as a scent and has now become fashionable. This new-found appreciation is based on an aromatic sophistication that can hold its own against all comers. Above all, sage has the advantage of avoiding the connotations of laundries and detergent that have stuck to lavender. A nose that comes across clary sage for the first time can therefore capture all its nuances and virtues without being led astray by such trivial distractions.

Genus and species: Salvia sclarea
Family: Lamiaceae
Type: Herbaceous plant (0.4–1 m)
Main growing areas: France, Hungary, Russia, USA, China
Harvest: Late June–late July
Parts used: Stem and flower
Treatment: Steam distillation (essence) and extraction with solvents (absolute)
Olfactive family: Aromatic
Facets: Green, herbaceous, clean, woody, camphorous, amber

The Collection

Masculine

L'immensité

The blue of the sea
Has no limits
And the heart, new vigour.

Composition
Ginger
Clary sage
Rock rose

Nouveau monde

In the chiaroscuro of the world
The explorer
Is also looking for love.

Composition
Cocoa
Saffron
Oud Assam

Imagination

Dream harder
See the big picture
Boldness is brilliant.

Composition
Citron
Neroli
Guaiac wood

Orage

And the world is reborn
In the imprint of the sun
In a passing shower.

Composition
Iris
Patchouli
Vetiver

Au hasard

Cloudless sky
The journey is unbounded
A smile in the wind.

Composition
Bergamot
Freesia
Sandalwood

Météore

Strength and light
The heady energy
Of seduction.

Composition
Mandarin
Neroli
Vetiver

Sur la route

Her breath
On the path of her skin
Sweetness without limit.

Composition
Bergamot
Nutmeg
Peru balsam

Mille feux

The grain of her skin,
Incandescent firework
Of love.

Composition
Bergamot
Osmanthus
Sandalwood

Contre moi

How many nights
Left to live love
On the other side of the world?

Composition
Bergamot
Orange blossom
Bourbon vanilla

Attrape-rêves

But the night lingers on
In the festive spirit
Infinite tenderness.

Composition
Cocoa
Peony
Patchouli

Matière noire

In the black of space
The luminous halo
Of a comet, a festive heart.

Composition
Blackcurrant bud
Narcissus
Patchouli

Le jour se lève

Innocent promise
Of a triumphant morning
Like the beginning.

Composition
Mandarin
Arabian jasmine
Osmanthus

Turbulences

Season of love
The bees kiss
The hearts of the flowers.

Composition
Jasmine
Tuberose
Sandalwood

Dans la peau

Fiery sun
A few burning kisses
On your skin.

Composition
Apricot
Magnolia
Natural leather

Feminine

Rose des vents

The breath of petals
Of a wild rose
Heckled by the wind.

Composition
Blackcurrant bud
Centifolia rose
Violet leaf

Apogée

In the early hours
Playing to get lucky
In all innocence.

Composition
Tangerine
Lily of the valley
Guaiac wood

Étoile filante

The velvet night
Lights up with hope
Make a wish, quick!

Composition
Osmanthus
Jasmine
Magnolia

Spell On You

A love song
Touching the nerve
Incurable melody.

Composition
Iris
Arabian jasmine
Acacia

Heures d'absence

The cool morning air
Carries the flowery wind
Back to Grasse.

Composition
Arabian jasmine
Mimosa
Sandalwood

Cœur battant

The vital pulse
Of euphoric nature
In contagious exhilaration.

Composition
Ambrette seed
Jasmine
Patchouli

Oriental

Les sables roses

In the pink dawn,
The dunes and the sky
Fuse their differences.

Composition
Oud Assam
Rose
Ambergris

Ombre nomade

With the march of the sun
The dunes come to life
The euphoria of a mirage.

Composition
Oud Assam
Incense
Benzoin

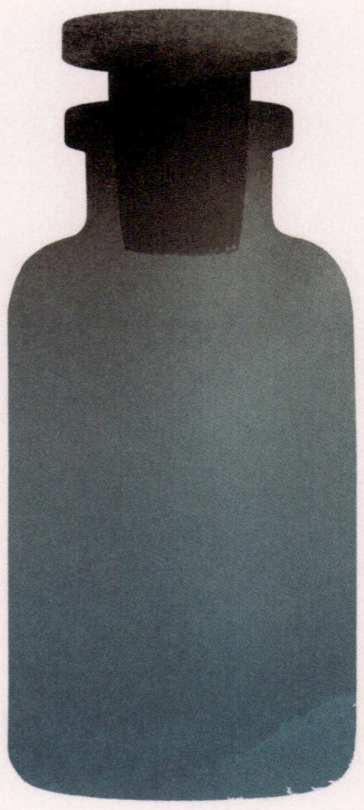

Nuit de feu

The dancing flames
Of a desert fire
A few wafts of eternity.

Composition
Incense
Natural leather
Oud Assam

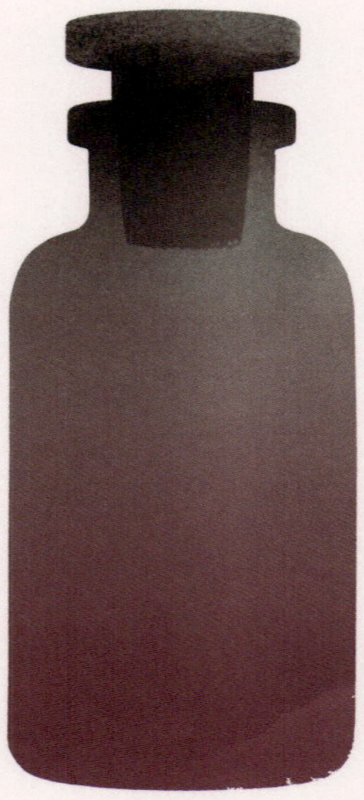

Fleur du désert

Out of the arid sand
A flower arises
A dazzling mirage.

Composition
Centifolia rose
Arabian jasmine
Oud Assam

Pur oud

Everything goes right
In this eternity
Of nature.

Composition
Oud Assam
Musk

Cologne

California Dream

The high-wire sun
Scatters the ocean
With peaceful lights.

Composition
Mandarin
Ambrette seed
Benzoin

Cactus Garden

Under a verdant arch
A patio of lushness.
Breathe! Breathe!

Composition
Mate
Bergamot
Lemongrass

On the Beach

The ecstasy of the body,
Drained by the sun
Languishing on the sand.

Composition
Yuzu
Neroli
Cypress

Afternoon Swim

Diving into the blue
The regenerated body
Revels in the summer.

Composition
Mandarin
Orange
Bergamot

City of Stars

The fluorescent night
Electrifies the skin
With citrus and freedom.

Composition
Blood orange
Tiare flower
Sandalwood

Sun Song

The watery orange sun
In the first gulps
Of summer.

Composition
Citron
Orange blossom
Lemon

Pacific Chill

Before the undulating dance
Of the waves
The heart is free.

Composition
Blackcurrant bud
Carrot seed
Aromatic herbs

Les Extraits

Dancing Blossom

Flight of petals
The garden dances
The world turned upside down.

Composition
Arabian jasmine
Centifolia rose
Osmanthus

Cosmic Cloud

In the nocturnal meadows
Of space
Gathering starry flowers.

Composition
Blackcurrant bud
Ambrette seed
Tonka bean

Rhapsody

The space of a moment
Restarting its life
Curious alchemy!

Composition
Ylang-ylang
Mate
Vetiver

Symphony

A dazzling pool
Of golden sunshine
The fiesta of the world.

Composition
Grapefruit
Bergamot
Ginger

Stellar Times

Sublime, subliminal
Journey to the heart of time
It is the hour of stars.

Composition
Orange blossom
Peru balsam
White amber

Myriad

In the dark immensity
A ray of gold
Lights up hope.

Composition
Cocoa
Centifolia rose
Oud Assam

Louis Vuitton
in a State of Grasse

Terroir. Behind this simple word, which shares the same etymology as – and is often confused with – 'territory', behind this word considered untranslatable into any other language, lies a unique concept and a reality that is utterly French (even neighbouring countries like Spain, Italy and Germany refer only to 'territory'). With regard to terroir, the Larousse dictionary of the countryside speaks of 'agricultural capacities due specifically to the nature of the soil, the local microclimate and the lie of the land'. In fact, the word is far richer and more complex. Although it does indeed cover the soil in which a plant grows – and, therefore, other considerations such as geography, orientation, geology and climate – 'terroir' also refers to the know-how of the people who cultivate, harvest and transform a plant. Use of the word was long confined to viticulture, but it has now been extended to the cultivation of plants for perfume, alluding to the mysterious alchemy of an alliance between human beings and the land.

Jacques Cavallier-Belletrud has been a champion of this ethic of terroir for many years. 'We have nothing to do with folklore. If the perfume industry became established between the sea and the Prealps, it was to take advantage of this climate of medium altitudes, neither too hot nor too cold, which makes plants more fragrant. Our predecessors totally understood this!' The territory of Grasse may be restricted to a few hundred hectares, but it has been blessed by the gods, allowing flowers to fulfil their olfactory potential and represent an alternative to globalization.

It was this predilection for the terroir that explained why the creative workshop was set up in the Fontaines Parfumées, in the heart of the city of Grasse, and not somewhere else. This place provides both ancestral knowledge and the people capable of putting it into practice and bringing it to fruition. The truth is, perfume runs through the veins of everybody in Grasse, whether they are the son of a perfumer, the granddaughter of a distiller or the great-granddaughter of a rose or mimosa grower. It is usually reckoned that 15,000 of the city's 51,000 inhabitants earn their living from the perfume business, either directly or indirectly. In fact, the entire population has been devoted to perfume for 400 years and this cultivation of raw materials has infused the entire landscape.

Human beings have undoubtedly made themselves an integral part of the concept of terroir. Indeed, the concept of 'country' cannot be expressed without them. Just as music remains silent without a player, this terroir would be bleakly quiet without producers of perfume plants, without distillers and without perfumers.

Ever since Louis Vuitton was established in Grasse in 2016 and signed contracts with farmers and distillers, local perfume plants have ridden a new wave of popularity, despite their high price. Ingredients from Grasse are twenty to twenty-five times more expensive than their equivalents from elsewhere. Louis Vuitton has decided to make

pure luxury available to its master perfumer, not out of self-aggrandisement but simply because the nose – to use his industry title – can tell the difference.

If a flower is truly rich and fragrant, the first reason is the exceptional terroir in which it blossomed. As the popular saying affirms, 'Wine always smells of its terroir'. 'Our Romanée-Conti is jasmine from Grasse – it's incomparably beautiful!' exclaims Jacques Cavallier-Belletrud. There is nothing old-fashioned or retrograde in this approach, and there is absolutely no desire to create a museum. It is precisely because this traditional know-how has been acquired and assimilated that Louis Vuitton's teams are able to work with natural ingredients in their own way, subject them to other treatments and other extraction techniques, and come up with, for example, an 'autumn jasmine', a sort of vintage raw material obtained from jasmine at the end of the harvest. In order to more fully express the richness of flowers from the terroir, Jacques Cavallier-Belletrud joined forces with Swiss company Firmenich to develop a revolutionary technique: using supercritical CO_2 extraction on fresh flowers. Up till then, this method, which shows the utmost respect for plants, had been successfully used only on dried materials. Jacques Cavallier-Belletrud was overjoyed with the results: 'The great difference is that by working at just 25 °C, the liquid absorbs all the fragrant molecules through immersion over a precise period of time, producing a 100 per cent natural flower extract, without any solvents derived from petroleum.' Louis Vuitton has obtained exclusive rights to this technology from Firmenich for all the flowers and materials produced in the Grasse terroir, starting with two of its unsurpassable emblems: May rose (*Rosa x centifolia*), found in the Louis Vuitton perfume *Rose des vents*, and jasmine (*Jasminum grandiflorum*), much of which is harvested to make *Étoile filante*.

The inclusion in 2018 of Grasse and its 'skills related to perfume' on UNESCO's Intangible Cultural Heritage of Humanity list has only strengthened a movement that has been active for several decades to increase the profile of this terroir. At the moment, Louis Vuitton's perfumer is alone among his contemporaries in the major luxury firms in settling in Grasse – a move that represents a way of drawing closer to its ancestral knowledge.

The general public may have already heard of *centifolia* roses and jasmine from Grasse, but it is less well known that Grasse is also the primary hub for the transformation of raw materials. 'The genius of Grasse is that it has managed to split the essence of patchouli – which comes from Indonesia – in order to retain its most beautiful part and then succeed in modernizing it,' explains Jacques Cavallier-Belletrud.

The latest stage in the movement to increase the profile of an exceptional terroir is the emergence of a 'Grasse' designation. The geographical indication Absolue Pays de Grasse ('Local absolute from Grasse') has just been granted by the French National

Institute of Intellectual Property (INPI) to seven companies from the region, including several suppliers to Louis Vuitton: Firmenich, Payan Bertrand and Robertet. This designation is distinguished by Grasse's unique capacity to extract biomass to make extraordinarily subtle essences. Just as there is Azores honey and Guérande salt, so there are Grasse absolutes of rose and jasmine – all of them products of great excellence. And these are evidence of the specific characteristics of the Grasse terroir that Louis Vuitton has drawn on to produce perfumes that are more sensitive, more sensual – in short, more alive.

To detractors who argue that such notions are both poetic and abstract, and that an uninitiated nose cannot distinguish a Grasse rose from a Bulgarian one, or a patchouli from Malaysia from one from Indonesia, Louis Vuitton responds that beauty requires an initiation process, and that, furthermore, a luxury firm has a responsibility to elucidate these concepts, to make them sufficiently concrete and tangible, to instil culture and knowledge in its customers and educate them about an ingredient's typology and geographical origin. The concept of terroir is neither a fable nor a myth – and in the perfumes of Louis Vuitton it is a reality that can be smelt on the skin and reverberates like the pulse of life.

The Fontaines Parfumées

From the beginning, Louis Vuitton chose to establish its creative activities in all its fields of expertise right in the heart of the region boasting the greatest experience and the most qualified craftspeople. In terms of perfume, Grasse was indisputably the centre of the world, the axis of gravity, the exact place to fulfil high ambitions and obtain the finest ingredients. A peaceful setting was required for Jacques Cavallier-Belletrud, so that he could create in a tranquil atmosphere. This disused *bastide* (country house), tucked away in a quiet, attractive spot far from the city centre's hubbub, fitted the bill perfectly. It was built in 1640, but in 1840 it was taken over by the Alziary company, which installed a workshop for processing raw materials for perfumes via distillation and enfleurage. In the 1920s, Émile-André Pélissier bought the building and established his own brand of perfumed water: Les Fontaines Parfumées. This visionary marketed water from the estate by proclaiming it to be 'naturally perfumed' – a legend so compelling that wealthy denizens of Grasse flocked to fill their flacons at the round Art Deco fountain, in which a few canisters of perfume had been judiciously concealed. This business subsequently dwindled, and eventually ceased altogether in the 1970s. The house remained abandoned for almost fifty years, a sleeping beauty almost forgotten by the locals – apart from an occasional nostalgic glance – before being bought by Louis Vuitton. The company was driven by a desire to establish its creative hub there and turn it into a showcase perfumery with exacting standards that proclaimed loud and clear its love for locally sourced raw materials and the skills that go hand in hand with them.

After four years of studies conducted in conjunction with Architectes des Bâtiments de France (an official organization overseeing French historic buildings), almost 200 artisans and their teams, mainly based in the region, were commissioned to restore the building, rigorously respecting local expertise. An immaculate English-style lawn evokes a heavenly sense of freshness in front of the ochre walls of the large house. To one side, a bed displays 350 mainly native essences, while there are also several varieties of citrus fruit from all over the world and some twenty types of mint, not to mention a field of *centifolia* roses here, another of *grandiflorum* jasmine there... All in all, a magnificent swirl of smells that swings like a garden full of jazz! The perfumer has only to stroll down the long flight of stairs to breathe in the soul of Grasse and imbibe its truths among these fragrant plants. He can forget about their stuffy formal names and allow himself to be swept away by their scents and wait for the muse of inspiration. It took almost four years of preparations to make this olfactory symphony ready for performance. This garden of scents owes its existence to another local man: the landscape architect Jean Mus, a poet of plants and arbours. He created the most beautiful atlas of scents and finally fulfilled his dream of creating a *kaiyū-shiki-teien*, a Japanese

strolling garden, in Provence. It may lack a pond, but there is a touch of water, in the form of a tinkling fountain fed by the providential Foux spring, which was used to water fields of flowers and tan leather in the sixteenth century.

A Tuscan melancholy is gradually taking over this secluded Eden, fostered by the outlines of the villa. Amid its verdant surroundings, the house stands tall in its smiling austerity. On the ground floor, there are a few cool rooms with dimensions typical of the period: a sitting room, a library where guests can wait, a convivial kitchen. A large staircase climbs towards the blue Mediterranean skies. Jacques Cavallier-Belletrud has his office on the first floor: a cosy den piled high with books, magazines and works of art, an incubator for olfactive projects of all kinds, whose only trace is the dozens of scented testing strips scattered around. On the top floor, there is a futuristic workshop with glass walls, like the bridge of a ship at anchor, where the creator can translate even his wildest dreams into smells.

Essentially, the Fontaines Parfumées is an unpretentious family house, with just enough comfort and refinement to dispel any desire to leave. It is the ideal place to build up an unorthodox perfumery freed from both rigid codes and passing fashions. Behind the high wrought-iron railings, it is not merely a Grasse manor house that has been reborn: perfumery itself has been given a new lease of life.

Jacques Cavallier-Belletrud:
The Master of Composition

Jacques Cavallier-Belletrud is a connoisseur of scents with a nose fine-tuned to recognize the tiniest component of an essential oil of patchouli or absolute of jasmine. He has inherited from Grasse, his cultural cocoon, a sensual, intuitive knowledge of natural raw materials from far and wide. Vetiver, Damask rose, rock rose, geranium – he knows them all intimately because he has seen them in the ground, after the harvest and during extraction. He can elucidate the botanical origin of a mimosa or a rock rose, as well as the characteristics of their florescence, the harvesting ritual surrounding them and the extraction process that draws out their quintessence.

He generously likes to give each of these wonderful essences the chance to shine by putting them into the spotlight, so that they can be seen – or, rather, smelt. What is the point of using a sublime essential oil of eastern red cedar if it is not noticeable in the composition?

Jacques Cavallier-Belletrud is not interested in modest flowers: a bouquet has to be generous, lush and profligate, and it has to produce a physical experience that takes it beyond an intellectual exercise. Although every essence carries within it a cultural component that inevitably enters into any composition, for this indefatigable inventor the creative act involves deviating the natural product from its initial vocation. Jacques Cavallier-Belletrud never confines an extract to any previous evidence of its function, and he pushes this philosophy to its limits: 'There is no benefit in having the finest raw materials in the world at your disposition if you're always going to make them say the same thing'.

In the woody landscape of *Orage*, he decided to make a rose blossom, using it not for its florality but rather for its freshness. Curiously, the 'flower of flowers' is not seen here as simply one of them: its face and its nature are changed simply because the perfumer chose to change the way he looked at it, thereby modifying our own perception as well.

Jacques Cavallier-Belletrud loves to explain, with his mellifluous voice and charismatic raconteur's brio, how materials establish a conversation and provide answers. He has a craving for this type of knowledge, but it is not an end in itself – otherwise he could have ended up trapped by it, in an old-fashioned perfumery more like a museum. Not that he has anything against tradition – just as long as it is not used as to cover up tics and manias. He waxes lyrical over a touch of tonka bean that has been left to soak overnight and already gives off something entirely different from before. 'One can feel that there is life there… It is the genius of nature,' he explains. And that genius combines with the genius of the man who extracted it to come up with a product that is even more beautiful. Jacques Cavallier-Belletrud knows all this, but he wants to go even further. He is on the lookout for natural ingredients that will spring a surprise on him.

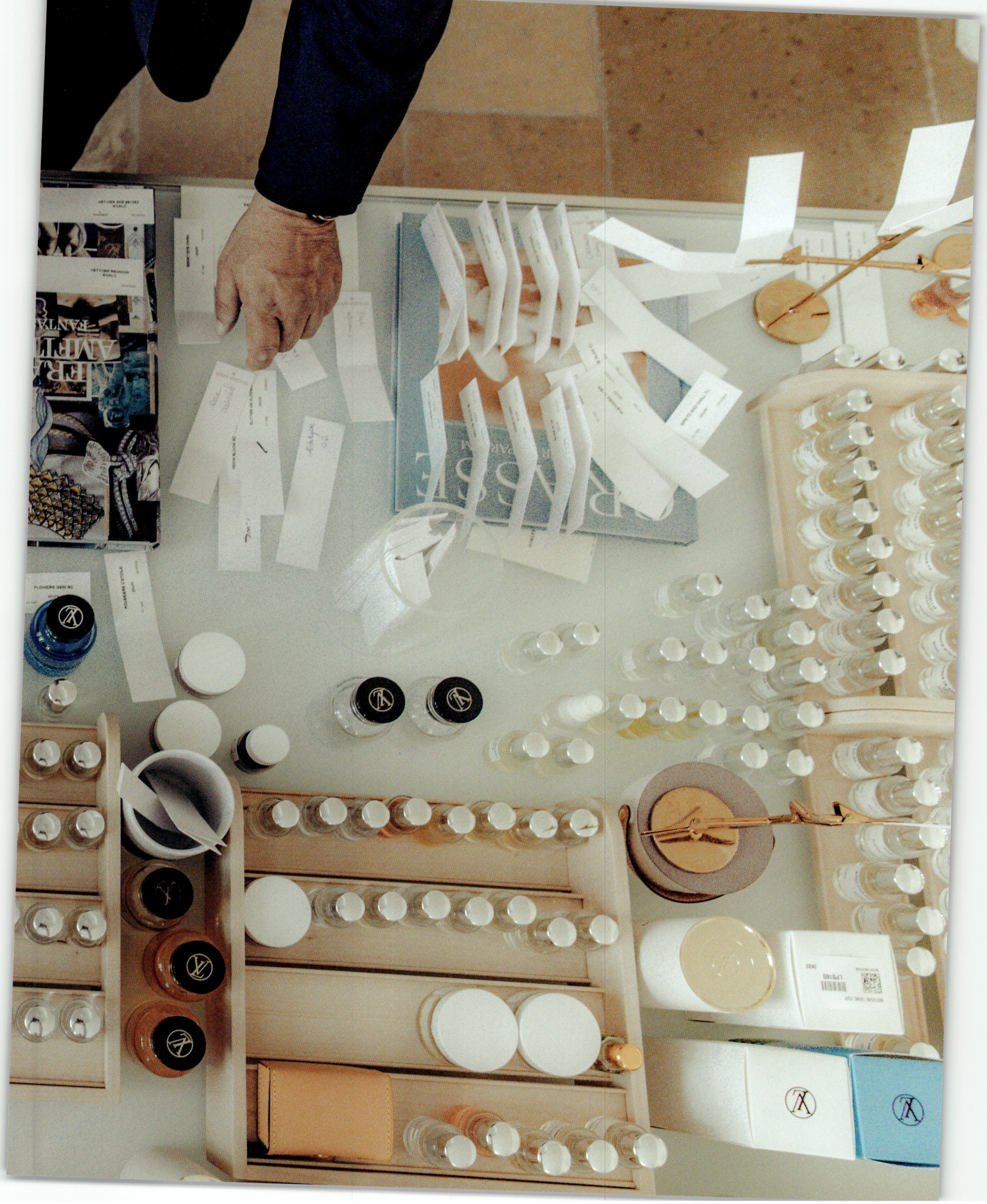

METTLER TOLEDO

TOLEDO

Another element that comes into play is seasonality: bergamot is not the same, for example, at the beginning and at the end of the harvest, and a perfumer can opt for one or the other. This explains why his work never repeats itself or loses its bearings, and also why he has never been drawn towards the folkloric or the picturesque.

Many perfumers are wary of natural materials, which can quickly get out of control and overwhelm a creator who has not learned how to master it. Jacques Cavallier-Belletrud understands, however, that he owes everything to such raw materials, as their very versatility dispels the boredom of routines and the conformity of rules and dogmas. They never fail to transport him – preferably somewhere unexpected – and accordingly he never exhausts their possibilities and potential combinations. He sees a natural resource as a wellspring of opportunities that lead straight to a world of beauty and happiness. Each extract displays several characteristics, and a perfumer traditionally 'plays' the main, most obvious one while also choosing from among its other, less frequented byways. Jacques Cavallier-Belletrud likes to give the example of one of his favourite flowers, the Chinese osmanthus, which carries a delicious note of apricot but can also express a smell of skin – especially at high doses – that utterly changes the picture. He loves a natural ingredient to such a degree that he is even capable of backing it into a corner – all in the good cause of coaxing it to say something that it has never dared say before. This is how he came to develop an infusion of leather based on naturally tanned leather offcuts. A traditional process, but a new idea. It is always the raw material that inspires him. After that, the road is open…

By dint of sheer enthusiasm, Jacques Cavallier-Belletrud has given Louis Vuitton a perfumery devoted to pleasure whose clarity is matched only by its harmony. A collection of compositions shot through with visions, with everything perfectly in place. There is a distinctive light that penetrates each perfume, but the raw material provides the key, ultimately giving the impression that you can touch it with your finger and caress it. We now say that something is 'Vuitton', just as we would say that something is 'cashmere' or 'raw silk'. But make no mistake: this apparent formal facility and unswerving clarity are the fruits of a solid, geometrical, considered construction. The perfumes of Jacques Cavallier-Belletrud are like houses with solid foundations… Never before have rigour and fantasy cohabited so perfectly.

NUIT DE FEU LOUIS VUITTON
FLEUR DU DÉSERT LOUIS VUITTON
PUR OUD LOUIS VUITTON
L'IMMENSITÉ LOUIS VUITTON
NOUVEAU MONDE LOUIS VUITTON
IRIS DE FLORENCE Absolue
JASMIN DE GRASSE CO₂ Absolue
JASMIN D'EGYPTE Absolue
JASMIN SAMBAC DE CHINE Absolue
MAGNOLIA DE CHINE Essence
TUBÉREUSE D'INDE Absolue

L'IMMENSITÉ
LOUIS VUITTON

Index of Raw Materials

Absolute

A plant extract obtained with volatile solvents. This method involves macerating plants in a tank and washing them several times, in order to release their fragrant molecules. The solvent is then boiled to produce a concrete (in the case of flowers) or a resinoid (using other parts of the plant). The concrete is washed and chilled several times to obtain the product's quintessence – the absolute.

Alcohol

Solvent used to dilute the fragrant concentrate. Perfumes are generally a minimum of 80 per cent alcohol. Ethanol (also called ethyl alcohol) is widely used because it is a neutral substance that does not alter the scent of fragrant materials.

Alcoholate

A liquid obtained by macerating plant substances in alcohol prior to its subsequent distillation.

Aromates

Fresh, invigorating raw materials, often associated with rituals of hygiene and grooming. They usually incorporate lavender, mint or aniseed.

Balsam

Resin taken from an incision in a tree trunk and then distilled with steam. The main balsams in a perfumer's repertoire are benzoin and the balsams of Peru, copaiba and tolu.

Balsamic

A sweet warm, soothing fragrance evoking raw materials described as 'balmy'.

Biomass

A term designating the raw material from a plant (flower, leaf, seed, resin) prior to the extraction process.

Chilling

An operation involving the cooling of an alcoholic solution to facilitate the precipitation of its less soluble substances (plant waxes) and obtain a product that is not cloudy after it is filtered.

Chromatography

A physical-chemical method used to identify the molecules involved in the composition of a scent, along with their proportions, in order to define its structure and, if required, reproduce it.

Cold pressing

The action of mechanically pressing the zest of citrus fruits to extract its essence, without applying heat. The essential oil is then separated from the water content with a centrifuge.

Communelle

A set of batches of natural essences or other raw materials designed to achieve a constant quality, from one harvest to another, for example.

Composition

The result of the combination of the different raw materials used in a perfume.

Concentrate

The perfumed composition prior to being mixed with alcohol. The subsequent mixture with alcohol at different dilutions will produce an eau de Cologne, an eau de toilette, an eau de parfum, a perfume or a perfume extract.

Concrete

Perfume in solid form. This more or less hard paste is obtained by solvent extraction of fresh natural raw materials, such as flowers, leaves, etc.

Enfleurage

A traditional method that takes advantage of the capacity of fatty substances to naturally absorb scents. The fat can be hot or cold, depending on the resistance of the plants. In the case of hot enfleurage, plants are macerated in fats that have previously been heated, until their essence is obtained. This fatty mass, known as a pomade, is then placed in a mixer with the alcohol that will absorb its scent. It is filtered several times, before finally yielding the absolute. Cold enfleurage is used to process very delicate flowers; their petals are laid on a thin layer of cold fat (traditionally animal fat, but nowadays vegetal fats are preferred).

Essence

This originates from citrus fruits and is obtained by cold expression of the small vesicles in the peel. Essence is generally sensitive to light.

Essential oil

A raw material produced by an aromatic plant, obtained through distillation or cold pressing. It provides a perfume's most volatile notes, known in the trade as 'top notes'.

Extraction

The removal of different components from a single raw material and their separation via the various methods applied in perfumery: enfleurage, distillation and extraction with solvents or CO_2.

Extractor

A large steel receptacle resembling a submarine in which a plant is placed before being covered by a solvent, which becomes impregnated with its scent after several successive washes.

Fractionation, fractional distillation

A selective distillation, also known as 'rectification', which makes it possible to recover various components of a product by playing with their boiling points. The different fractions are analysed and put together; this process also permits the elimination of undesirable elements, such as musty or camphorous notes of patchouli. The resulting extract is called a 'heart'.

Headspace

This technology, which emerged in the 1970s, makes it possible to use a neutral gas to capture the fragrant molecules of a living organism or an object. A perfumer can reproduce a scent after undertaking a chromatographic analysis.

Hesperidia

Citrus fruits (citron, bergamot, mandarin, grapefruit, etc.), which produce light, volatile, tonic notes conveying a sense of exuberance.

Hydrodistillation

Extraction method invented by the Persians and first recorded in 961 CE. It requires a copper still in which petals are placed, before being covered with water, which is then brought to the boil. The resulting steam conveying fragrant materials is then cooled, becoming perfumed water and being directed into a receptacle. The water and the essential oil are easily separated owing to the density of the oil.

Hydrolate

Once a flower has been distilled, a perfumer is left with, on the one hand, essential oil and, on the other, the water derived from steam during the decantation, known as the hydrolate. This floral water is imbued with the same aromatic components as the essential oil, but in a lower concentration.

Infusion

An operation that involves putting a solid or paste-like raw material in prolonged contact with ethyl alcohol (around six months). The result was traditionally referred to as a tincture. Louis Vuitton's perfumer has used this method to create an unprecedented and exclusive infusion of leather.

Isolate

An isolated component of a natural raw material (for example the geraniol found in roses), which can be extracted by fractioning the essential oil into its various components.

Maceration

The final stage in a perfume's manufacturing process, prior to bottling. It involves leaving the mixture of concentrate and alcohol to rest, while cold, for up to four weeks, to enable it to acquire its definitive identity. Maceration plays a decisive role in emphasizing particular facets of the composition.

Maturation

Once the formula for a perfume has been composed and weighed to ensure that all its notes combine harmoniously, the maturation process takes its course: the perfume's concentrate is left to rest for a few weeks before being mixed with an excipient (denatured alcohol specially designed for perfumes).

Mute flowers

A flower is considered 'mute' when its fragrance cannot be captured. This is particularly true of lily of the valley, lilac and violet. In such cases, perfumers can reconstruct the scent of a mute flower with synthetic molecules, some of which can be derived from nature.

Naturals

Raw materials of natural origin derived from flowers, fruit, leaves, spices, rhizomes, bark, wood and resins. There are around 500 naturals in all.

Petitgrain

Essential oil produced by the distillation of a citrus tree's leaves and twigs.

Pomade

Fatty perfumed mass obtained from flowers through enfleurage.

Raw material

An initial product, of natural or chemical origin, used to manufacture a perfume.

Rectification

The removal of a raw material's impurities through distillation.

Resinoid

An extract from a dry material (balsam, gum or resin) obtained via treatment with volatile solvents.

Solvent

This serves to dissolve a plant's fragrant material without modifying it. Hexane is used for fresh raw materials such as flowers, while ethanol is preferred for the treatment of dry raw materials.

Sourcer

A globetrotting seeker of fragrances, on the lookout for a new raw material or a new variety of an already established ingredient. A sourcer can sometimes act as a buyer for a producer.

Sponge extraction

Traditionally, extraction was done by hand. The citrus fruit was cut in half, emptied of its flesh, and its rind moistened. After a short interval, the essence was squeezed out of the fruit on a set of sponges. After decantation, the essence of the fruit was collected in a vase by simply pressing the sponges. This traditional method has been rediscovered in recent years.

Supercritical CO_2

This extraction technique has been used in perfumery since the 1990s (CO_2 extraction of pink peppercorn dates back to 1993) and provides an end product of exceptional purity. It involves compressing carbon dioxide gas until it turns into a liquid. In this 'supercritical' state, this liquid acts as a solvent and draws out a plant's scented components. Because the temperature remains relatively low, the plant does not deteriorate. This technique is particularly useful for raw materials that are very delicate or have little fragrance: wood, bark, seeds and spices.

Jacques Cavallier-Belletrud would like to thank Pietro Beccari, Michael Burke, Stefano Cantino and Camille Cavallier, as well as Claire Brun, Isabelle des Garets, Mélissa Giraud, Julien Guerrier, Aurore de la Morinerie, Yann Musquin, Timothée Nicolas, Lionel Paillès, Dominique Roques, Axelle Thomas, Anthony Vessot and Sébastien Zanella.

Lionel Paillès would particularly like to thank Jacques Cavallier-Belletrud, who welcomed him so often to Grasse and vividly conveyed both his love for raw materials and the excitement of the demanding world of perfumery; Cathy Délia for her warm welcome at the Fontaines Parfumées and her innate sense of organization; Mélissa Giraud and Camille Cavallier for their generosity in sharing their passion for fragrance; Julien Guerrier for allowing him to fulfil a dream and kindly overseeing the publication of this book; Anthony Vessot for his endless patience, his good humour and his attentive and enlightening reading; Axelle Thomas for her valuable help in the final stages of this long project; and all the cultivators of perfume plants, in Grasse and elsewhere, who shared their botanical and agricultural knowledge.

Aurore de la Morinerie would like to thank Jacques, Julien, Lionel, Anthony, Raphaëlle and Isabelle.

Thanks also to Frédéric Badie, Pietro Beccari, Marie-Eugénie Bouge, Claire Brun, Stefano Cantino, Valérian Charmasson, Mathias Clottu, Roxane Cornet, Cathy Délia, Isabelle des Garets, Julien Guerrier, Léa Guillon, Virginie Hairanian, Louise Husson, Édith Jarboua, Anaëlle Lecointe, Laura Lingibé, Patrick Mauriès, Carène Monnet, Yann Musquin, Timothée Nicolas, Vincent Proal, Lewis Ronald, Dominique Roques, Ivana Simic, Axelle Thomas, Cynthia van Moer, Anthony Vessot, Valérie Viscardi and Sébastien Zanella; to the team at Thames & Hudson in London: Julie Bosser, Paul Hammond, Susanna Ingram, Corinna Parker, Jill Phythian, Mark Ralph, Adélia Sabatini, Flora Spiegel and Clare Turner; and to the translators, proofreaders and picture researchers: Hélène Boraz, Matthew Clarke, Guillaume Müller-Labé, Sylvie Philippon, Kirsty Seymour-Ure, Jo Walton and Howard Watson.

Text translated from the French by Matthew Clarke

First published in the United Kingdom in 2024
by Thames & Hudson Ltd, 181A High Holborn,
London WC1V 7QX

First published in the United States of America
in 2024 by Thames & Hudson Inc., 500 Fifth Avenue,
New York, New York 10110

Designed by Studio Mathias Clottu

British Library Cataloguing-in-Publication Data
A catalogue record for this book is available from
the British Library

Library of Congress Control Number 2021950032

ISBN 978-0-500-02238-2

Printed and bound in Italy by Verona Libri

MIX
Paper | Supporting
responsible forestry
FSC® C119614